GENE WARS

GENE WARS

MILITARY CONTROL OVER THE NEW GENETIC TECHNOLOGIES

CHARLES PILLER and
KEITH R. YAMAMOTO

Library of Congress Cataloging-in-Publication Data

Piller, Charles.
Gene wars : military control over the new genetic technologies / Charles Piller and Keith R. Yamamoto.
p. cm.
Bibliography: p.
Includes index.
ISBN 0-688-07050-7
1. Biological warfare—United States. 2. Chemical warfare—United States. 3. Genetic engineering industry—Military aspects—United States. 4. Genetic engineering—United States. I. Yamamoto, Keith R. II. Title. III. Title: Military control over the new genetic technologies.
UG447.8.P54 1988
358'.38'0973—dc19 87-30629
CIP

Printed in the United States of America

First Edition

1 2 3 4 5 6 7 8 9 10

BOOK DESIGN BY RICHARD ORIOLO

The word "book" is said to derive from *boka*, or beech.
The beech tree has been the patron tree of writers since ancient times and represents the flowering of literature and knowledge.

ACKNOWLEDGMENTS

This book could not have been written without the help of many others. We are particularly grateful to Paul Hager, Jonathan King, Sheldon Krimsky, Rodney McElroy, and Barbara Rosenberg for their careful reading of and insightful comments on various sections of the manuscript. Their generosity of time and expertise proved invaluable to the final product. Their dedication to biological arms control was inspirational.

We are also indebted to David Castleman for his microscopic eye for bringing the best out in the manuscript; to Senator Alan Cranston and Ann Goldman of his staff for assistance in obtaining vital research materials; to Kathleen Rañeses for her expert, unflagging assistance with our research; and to Elysse Zelenko for helping us whittle down a ponderous mass of data to manageable proportions.

For helpful information or support at various stages of the effort, we also acknowledge Stanton Glantz, Donald Heyneman, Judy Klinman, Charles Lewis, Richard Novick, Ellin O'Leary, and Dale Van Atta.

For their belief in the importance of this issue, and their consistently valuable advice and encouragement, we thank our literary agents, John Brockman and Katinka Matson, and our editor and publisher, James Landis of Beech Tree Books.

Charles Piller and Keith R. Yamamoto
San Francisco, California

AUTHORS' NOTE

This book represents a five-year odyssey. I stumbled upon the potential applications of genetic engineering to biological warfare during research for an article on biotechnology in late 1982. Since that time the following friends, colleagues, and family members have nurtured my effort with timely suggestions, confidence in me, and belief in the importance of this work: Lynn Allen, Jim Bellows, James Bond, Diana and Kevin Bunnell, Shelly Coppock, Buck Cameron, Gary Fields, Liz Katz, Terry Leighton, Marion Nestle, Jeff Pector, Alice and Jack Piller (my parents), Geoff Piller, Steve Seimers, Norma Solarz, Chris Teidemann, Karen Teidemann, and Judy Timmel.

Special thanks also to the staff of *Synapse* at the University of California at San Francisco, for showing great flexibility when I most needed it; to the editors of the *Nation*, for having the foresight and confidence to publish prominently my early articles on biological warfare; to Michael Castleman, whose willingness to be a sounding board and consistently savvy comments made a major difference for the entire project; and most of all, to my wife, Surry Bunnell, for her enduring love and patience, constant encouragement, and incisive suggestions for the manuscript.

CHARLES PILLER

I was drawn into this project by my own strong feelings about the biological weapons issue and by my friend Charles Piller's gentle insistence that the perspective of a molecular biologist was needed to complement the historical, statistical, and political ramifications emergent from his years of investigation and analysis. I discovered a rather remarkable fit of our ideas and concerns and learned much in the process. As a scientist in whose laboratory the tools of genetic engineering are used daily to pursue fundamental biological questions, I hope that this book will be read by other scientists, for they hold a special opportunity and a special responsibility to influence the future of BW research. For nonscientists I have tried to clarify and demystify the new biological technologies and the policies that govern their use. These are crucial to the current BW debate, and it is now essential for all of us—scientists and politicians, nonscientists and nonpoliticians—to make informed choices about how those technologies will be applied.

KEITH R. YAMAMOTO

CONTENTS

CONTENTS

TABLES

1

INTRODUCTION: WAR AND THE REVOLUTION IN BIOLOGY

SCENES FROM THE NEW FRONTIER

In June 1989 the years of painstaking lab experiments finally pay off. Soviet molecular biologists at a top secret research institute in the Siberian city of Novosibirsk successfully employ genetic engineering techniques to create a novel disease organism—a superpathogenic strain of dengue fever, a viral disease endemic to tropical areas around the world.

The new organism is code-named D-7. A classified Soviet document (page 14), leaked by a dissident scientist, compares it with normal dengue.

D-7 is a marvel of sinister ingenuity, the perfect biological weapon (BW). At first glance it is indistinguishable from its natural parent, and even extended analysis writes off D-7 as the product of natural mutation. Rapacious and fast-acting, it reliably incapacitates everyone in its path, rendering people helpless but killing only a handful. More important,

	DENGUE	D-7
Mortality rate	0–1%	2–4%
Infectivity rate	75–95%	95–100%
Incubation period	3–15 days	2–4 days
Duration of extreme symptoms	2–4 days	7–15 days
Convalescence rate	Slow	Slow
Means of transmission	Aerosol or mosquito	Aerosol or mosquito

Novosibirsk has come up with a D-7 vaccine 100 percent effective in human trials.

The Soviets waste little time. By late July a refrigerated supply of the virus and several hand-held biological fogging devices are secretly shipped, via Cuba, to Salvadoran FMLN rebels deep in the hills of north-central El Salvador. At 2:00 A.M. on July 31 a small vaccinated commando squad creeps forward to the El Paraíso army base, a supply center for the northern region. They silently bathe the base in D-7. Before dawn their covert raid is completed, and the guerrillas steal back to mountain camps.

By the next afternoon a few soldiers return from patrol with the fatigue, joint pain, depression, and nausea characteristic of dengue—an unwelcome but familiar visitor. By the following morning more than 300 of the 800 men stationed at El Paraíso have symptoms, and most can barely lift themselves out of bed. The ranking officer, informing national commanders of the situation, is told to sit tight for a few days; by then the soldiers should start to recover, they say. By late evening patrols have been cut to a third their normal size.

At 2:30 A.M. on August 2 the rebels launch their most ambitious raid of the war. Local power lines are cut. Sections of the roads connecting the base to the capital, thirty-six miles to the south, are blocked. A strategic bridge halfway to San Salvador is blown up. A full-scale mortar attack immobilizes the small landing strip near the base and soon wears down the relative handful of government troops still unaffected by D-7. By dawn the unthinkable has occurred: Unable to retreat and without reinforcements, the troops and their arsenal fall intact into the hands of the FMLN. The battle takes fewer than two hours. The rebels call it a turning point in the civil war. . . .

By February 1990 the U.S. president is at his wit's end with Nicaragua's Sandinistas. They have enjoyed bumper coffee crops for three straight years. This has combined with record world coffee prices, because of consistent failures of the Brazilian harvest, to give the Nicaraguan economy its first hopeful signs since the revolution. After the Iran-contra arms scandal of 1986 the Nicaraguan rebels lost most congressional support, and their cause is nearly moribund. The president knows he has to stem the tide and avoid his predecessor's Central American failure as well as gain the initiative on foreign policy.

He doesn't realize help is on the way. Working on their own initiative, a handful of CIA operatives prevail upon the U.S. Army Medical Research and Development Command to solve the problem. The army's genetic engineers come up with an altered strain of coffee rust fungus—the organism that eliminated the coffee industry of Ceylon in the 1880s but had since been controlled with chemical fungicides.

Metal compounds, particularly copper derivatives, have been used to kill rust fungi for the past century. After a few months of genetic trial and error, army scientists create a new strain of coffee rust, producing a protein that binds tightly to copper ions, thereby neutralizing its effects, rendering the standard copper fungicides useless. An alternate fungicide—chemicals called carbamates—is still effective against coffee rust. It is not immediately available in Nicaragua, however, which has relied on copper for decades without incident.

In late June, in a covert action directed by the CIA, contra forces apply the genetically altered fungus to a major stretch of the Nicaraguan highlands crop. By the time Sandinista agricultural officials recognize that they need carbamates and are able to obtain supplies from East Germany to kill the creeping pest, 70 percent of the coffee has been destroyed. By October the economy is reeling. Forced to increase rationing of basic necessities, the Sandinistas face widespread discontent.

In desperation they seize the remaining large private coffee holdings that are not racked by the epidemic. This is the last straw for some prominent parliamentary opponents. Along with the plantation owners, they denounce the Sandinistas as incapable of managing the economy. They then defect to the contras, flee to Miami, and hold a press conference calling for renewed aid to overthrow the government. The president, buoyed by gains in the 1990 midterm elections, reexamines his options. . . .

Farfetched? Perhaps. But such scenarios are plausible to many American and Soviet leaders. Until now the use of biological weapons has been viewed as a minor issue effectively neutralized by the 1972 biological disarmament treaty. Conventional wisdom has held that such agents are dangerous to develop and handle, hard to disseminate effectively, and impossible to control. Widespread public revulsion against germ warfare has made the potential political and diplomatic risks associated with using these weapons unacceptably high. They have rarely seen action in modern military history.

This may soon change.

"The stunning advances over the last five to ten years in the field of biotechnology . . . mean more than new foods, pharmaceuticals and fertilizers. They mean new and better biological weapons. . . . It is now possible to synthesize BW agents tailored to military specifications. The technology that makes possible so called 'designer drugs' also makes possible designer BW," said Douglas J. Feith, deputy assistant secretary of defense for negotiations policy, in 1986 testimony before a congressional committee. "New agents can be produced in hours; antidotes may take years," he continued. "The major arms control implication of the new biotechnology is that the [BW treaty] must be recognized as critically deficient and unfixable."*

Feith's assertions are far from proved. In fact, some leading scientists and arms control experts consider the statements naive, irresponsible exaggerations. Does genetic engineering have the power to alter the fundamental strategic or tactical utility of BW? Not immediately. In the long term, though, the answer is less certain. But this may be the wrong question. The way the superpowers perceive BW is the best barometer of whether biotechnology will send the arms race off into a bizarre and dangerous spiral. And superpower perceptions have already recalculated the germ warfare equation.

BRAVE NEW WORLD

In November 1972, after a scientific conference in Hawaii, Herbert Boyer, of the University of California at San Francisco, and Stanley Cohen, of Stanford University, met at a Waikiki delicatessen for a late-night meal. Their freewheeling discussion over corned beef sandwiches led to a collaboration that was to forge scientific history.

The two were among a number of scientists pioneering experiments with DNA—deoxyribonucleic acid, a complex chemical that encodes each organism's hereditary information. DNA elegantly captures all the traits and characteristics—from eye color to Down's syndrome—that are passed on to future generations.

*Source notes begin on page 266.

Cohen and Boyer did not invent recombinant DNA (rDNA) technology: joining a part of the hereditary code of one organism to that of another. But they did establish a method of "gene splicing" that became a standard in molecular biology.

Genes, the basic hereditary units, are composed of DNA. The genes are strung together in long, continuous DNA strands—chromosomes—and sometimes small free-floating loops—plasmids. One can picture DNA as an architectural library whose thousands of volumes together constitute the complete blueprint for an organism.

A full complement of DNA—the entire genetic blueprint—is present in every cell of every living organism and directs each cell's activities. When a cell divides, it creates an exact replica of itself, including its DNA. The decoding of this genetic information yields specific products (proteins, hormones, sugars, or other substances) and is called "gene expression." The DNA code that controls gene expression is virtually the same for all organisms—from viruses to human beings. In this way the greatest and smallest of creatures are fundamentally related. This relationship is the cornerstone of rDNA technology.

Cohen and Boyer collaborated on a process that could biochemically "cut" a gene from the DNA of one organism, like a virus, and "splice" it into the DNA of the plasmid of another. The result: a chimeric molecule containing DNA from two species. The method allowed any gene from any species to be spliced into a plasmid.

To complete the process, scientists take advantage of the factory-like processes of the tiny cells. They return the rDNA into a rapidly proliferating host bacterium, where it is cloned—reproduced identically—each time the host cell divides. Since these cells can divide as often as once every half hour, a single cell can give rise to 5 sextillion (5 followed by 21 zeros) identical progeny in a single day. Through gene expression, the bacteria produce their normal products coded by their original complement of DNA. But they also produce the *viral* products coded by the virus gene spliced into the rDNA plasmid. (See Appendix 1 for further description of rDNA.)

This technology holds boundless medical promise. Cloning and other new biotechnologies can cheaply mass-produce highly refined natural products that the body synthesizes in only minute quantities. Shortages of human insulin, hormones, and immunity enhancers, such as interferon and interleukin, are becoming a thing of the past. The potential of rDNA for the development of vaccines dwarfs any previous breakthrough.

Beyond pharmaceuticals, rDNA has spawned a diversified, multi-billion-dollar industry, populated by venture capitalists, upbeat true-believer scientists, and multinational conglomerates. Their wares include pesticides, more robust crops with higher yields, and food additives, such as the sugar substitute aspertame.

Recombinant DNA played a major role in forging a new age in biology. "Biotechnology, broadly defined, includes any technique that uses living organisms (or parts of organisms) to make or modify products, improve plants or animals, or to develop microorganisms for specific uses," according to the congressional Office of Technology Assessment (OTA). Biotechnology per se is not new. "Since the dawn of civilization, people have deliberately selected organisms that improved agriculture, animal husbandry, baking, and brewing," the OTA notes. But "new biotechnologies," which have emerged since 1972, including rDNA and the examples below, have vastly expanded these abilities.

Hybridoma/Monoclonal Antibody Technology

When any foreign antigen—such as viruses, bacteria, or, in the case of hay fever sufferers, pollen grains—enters the body, the immune response works to attack and destroy the intruder. This natural defense system depends on antibodies—proteins that recognize and bind to the antigens. Each particular antibody associates with only one specific antigen, ignoring all others.

Until recently antibodies could be extracted only from blood serum. But serum extractions are an impure and dilute mixture of hundreds of different antibodies, each targeted to different antigens. In 1975, in a major biological breakthrough, hybridoma cells were created. These cells are unique for two reasons: They are "immortal"—that is, they can replicate indefinitely, unlike normal cells—and they produce a single, pure antibody in virtually unlimited quantities. These hybridoma products are called "monoclonal antibodies" (MCAs). MCAs have led to phenomenal advances in immunization, medical diagnosis, and vaccine purification. (See Appendix 1 for further description of how MCAs are produced.)

Bioprocess Technologies

Bioprocesses are systems that use living cells, such as yeasts and bacteria, to produce a desired organism or its chemical by-product. They are the original biotechnologies. Bioprocesses ferment alcohol, manufacture cheese and many other foods and beverages, and create a variety of industrial chemicals (such as acetone and butanol) and pharmaceutical products (such as antibiotics).

The role historically played by bioprocess technologies has been expanded vastly by rDNA, MCAs, and new methods of fermentation and cell culturing.

Comparing such techniques to an old bioprocess—such as brewing beer—is like pitting a supersonic jet against the Wright brothers' plane. At a fraction of the cost of old methods, new bioprocesses can efficiently produce a far purer and broader range of products than did their venerable forerunners. Combined with rDNA, these methods generate and refine substances never before available for practical use. The major limiting factor in the production of biological organisms and their chemical by-products is now storage capacity. And this is certainly not a large problem. Production plants that are tiny by conventional standards can mass-produce a desired product in record time, obviating the need for long-term storage in many cases.

TAKING A STEP BACK

From their inception rDNA and these other powerful new technologies inspired both awe and fear. Society's worst bogeymen soon sprang to life in science fiction and the media: wildly proliferating mutant organisms taking over the earth, wreaking havoc on the biosphere. Most scientists scorned these extreme visions. But they took safety questions seriously.

Following a meeting sponsored by the National Academy of Sciences in April 1974, scientists on the cutting edge of rDNA research took a step without precedent in science: They called for a voluntary halt to certain potentially dangerous rDNA experiments in order to evaluate the risks. A unique moratorium was established. It covered both rDNA manipulation of animal tumor viruses, and bacteria that involved toxins or the enhancement of resistance to antibiotics.

The following year 134 leading molecular biologists met at the Asilomar Conference Center near Monterey, California, to discuss the health, legal, and environmental implications of their work. Their recommendations led to the establishment of rDNA research guidelines by the National Institutes of Health (NIH).

These guidelines standardized both physical and biological containment. Physical containment involves safe lab practices and effective water and air filtration systems to prevent infection of lab workers or the release of recombinant organisms into the environment. Biological containment means using organisms that for their survival require growth media or environmental conditions not usually found in nature. "X1776," for example, is a "crippled" strain of *Escherichia (E.) coli.* In order to survive, it must be fed vitamin and amino acid supplements.

Recombinant DNA technology began by using only bacteria or yeast as hosts. But scientists can now produce genetically chimeric plants and animals. Inevitably this has spurred the most outrageous suggestions. The cloning of whole humans programmed for blind obedience by a totalitarian state is a well-circulated scare story. The proposition is considered ridiculous by anyone who understands the complexity of the genetic library.

But manipulation of the germ line—sperm, egg, or early embryo cells—is no fantasy. In 1982 researchers implanted the growth hormone genes of rats into mouse embryos. The result: mice that grew to twice their expected size. The implications of this work to agriculture are awesome: giant bovines that give the milk of three normal cows. And it holds great promise for easing human suffering. Thousands of genetically linked disorders, such as Tay-Sachs and sickle-cell anemia, could eventually be eliminated.

Theoretically the prospects are almost infinite. Picture parents preselecting their child's eye color. Imagine a genetic nose job. To be sure, these kinds of choices are unlikely to confront anyone for many years, if ever. But because we can reasonably contemplate them, even more profound questions are unavoidable. How far should humans go in manipulating our own genetic code? Who should be empowered to make such decisions? The scientific community? Legislators or government agencies? The general public?

Biotechnology's dizzying pace demands answers. But little societal consensus exists about how to find the solutions or even about how to educate the public to ensure broadly based, informed debate. As far as society stands from resolving these dilemmas, it is more distant still from understanding another potentially explosive implication of the new biotechnologies: biological and chemical warfare.

CHEMICAL AND BIOLOGICAL WARFARE DEFINED

This book is concerned with three types of unconventional weapons: biological weapons (BW), toxin weapons (TW), and chemical weapons (CW).

BW are "living organisms, whatever their nature . . . which are intended to cause disease or death in man, animals and plants, and which depend for their effects on the ability to multiply in the person, animal or plant attacked," according to a United Nations definition. The major forms of BW—each potentially deadly—are bacteria, viruses, rickettsia, and fungi. (For an overview of BW agents, see Appendix 2, Table 1a.)

Bacteria are nearly ubiquitous, single-celled microorganisms that can reproduce inside a host plant, animal, or person or independently, given a proper food supply. They can be killed by antibiotics (such as penicillin). Viruses are the simplest organisms—far tinier than bacteria and little more than a sheath of protein covering a scrap of genetic code. Viruses cannot reproduce independently. They are parasites, entering or injecting their genetic material into living cells.

Rickettsia fall between bacteria and viruses: large, complex and parasitic. Fungi are single- or multicellular plantlike organisms. Mushrooms are a subclass of fungi that show the incredible diversity of the genus—from benign taste treats to silent killers. Unlike green plants, which use chlorophyll to create their own food, fungi feed off other organisms or organic matter.

Chemical weapons are chemical substances in gaseous, liquid, or solid form that could be used with hostile intent because of their direct toxic effects on animals, plants, and human beings. Like all chemical agents, CW are inanimate and incapable of self-reproduction. Many potent industrial chemicals could be used, perhaps crudely, as weapons, but we are concerned only with the relatively limited number of chemicals that have strong practical potential in warfare, terrorism, or sabotage (see Appendix 2, Table 2a).

Nerve agents are the most powerful of these. They disrupt the transmission of nerve impulses in animals or humans, leading to a complete breakdown of all bodily functions. Death ultimately is caused by asphyxiation; the body fails to regulate the signals that instruct the diaphragm to pump the lungs. These are the primary CW agents stockpiled today.

Blister agents affect the lungs and create severe, painful blisters when they contact the skin. The most famous and widely used blister agent, mustard gas, caused hundreds of thousands of casualties during World War I. Lung irritants, such as chlorine, the original World War I gas, sear the lining of the lungs, leading to severe respiratory problems or death in high doses. Other respiratory irritants, such as tear gas, are widely employed in police actions and riot control.

Antiplant agents, or herbicides, are used for crop destruction and deforestation—to deprive an enemy of sustenance or hiding places. The infamous Agent Orange, used by the United States in Vietnam, is perhaps the classic wartime herbicide.

In general, toxins are poisons produced by living organisms—from shellfish to fungi (see Appendix 2, Tables 3a and 4a). They occupy a middle ground between chemicals and organisms. The active ingredient in poison mushrooms is a toxin. Botulinum toxin, a chief cause of food poisoning (botulism), is produced by a bacterium. Snake venom is also a toxin. Because many toxins have legitimate medical uses, their definition as a weapon is based on clear hostile intent.

Chemical and biological weapon agents are classed as harassing, incapacitating, or lethal. But these designations lie on a continuum based on environmental conditions, dosage, method of dissemination, and the variations of individual response. In sufficient concentration, any CBW agent can kill (see Appendix 2, Table 2a). Some agents cause unintended effects that cross over categories or defy precise definition. Agent Orange, for example, probably caused thousands of cancers although it was never directed against human beings.

APPLYING THE NEW BIOTECHNOLOGIES

The new biotechnologies promise to "improve" CBW agents and enhance their utility far beyond the capacity of classical biochemistry, both qualitatively and quantitatively. Whether this potential can be decisively exploited has yet to be determined in definitive examples whose results are available to the general public. But the following applications of the new biotechnologies have pushed CBW to the foreground of arms control concerns:

- Drug resistance. The genetic basis for bacterial resistance to antibiotics and viral resistance to other drugs is well understood. Genes that confer such resistance can be transferred to a BW agent to thwart medical countermeasures.
- Increased hardiness. Finding a way to keep aerosolized microorganisms from dying once they are sprayed from aircraft or exploded from bombs has been one of the most vexing questions for BW planners. Solar radiation, drying, and temperature fluctuation easily kill most agents adapted to live within humans or animals. But microencapsulation—a novel method of protecting individual BW organisms within

organic compounds—has already extended the range of agents that can be weaponized effectively.

- Defeating vaccines, natural resistance, and diagnosis. Our immune system's antibodies overcome a virus or other BW antigen by targeting the organism's specific surface structure. Using rDNA to make minute changes in this antigenic surface could render antibodies ineffective.

  Such methods could lend chilling realism to the Soviet dengue scenario. Scientists could learn which genes to alter in order to create a new surface structure, unrecognized by antibodies produced in people who have been infected by any of the naturally occurring strains. Virtually anyone exposed would contract the new disease. Through rDNA methods a form of the virus could be created that would frequently mutate—in essence making many "mistakes" as it self-replicates. This would lead to a disease of longer duration because the body's defenses would have to learn to recognize each of the various forms of the new dengue virus.

  Similarly, many diagnostic methods are based on the detection of certain sites on a disease organism's surface. Altering these sites could render a BW agent "invisible," thereby frustrating appropriate treatment.
- Unlimited vaccine development and new biodetection abilities. A nation probably would not use a BW agent unless it could protect its own people from infection. As Table 1a in Appendix 2 indicates, only a handful of organisms—for which vaccines or other proved protective measures exist—have ever been standardized and manufactured as weapons.

  Just as they have become tools to thwart vaccines, MCA and rDNA technologies have revolutionized vaccine research. Under modern bioprocess methods vaccines are simple to mass-produce, and in the foreseeable future they will be created for nearly all known potential BW agents. Meanwhile, MCA technology can detect the presence of biological agents in the environment with a sensitivity unthinkable a decade ago. These developments may reduce an aggressor's fears of backfire and retaliation.
- Increased virulence. Disease sympoms often stem directly from toxins secreted by a pathogen. Anthrax toxin, for example, is the active ingredient of the anthrax bacterium. The genes that regulate toxin production may be manipulated to enhance an organism's virulence. The result would be a more powerful, faster-acting, and invasive weapon, one that would infect and kill more reliably.
- Weaponization of innocuous organisms. Certain harmless micro-

organisms, such as *E. coli,* are a normal part of the body's ecology. By the transfer of genes that regulate the production of disease-causing toxins to these helpful microbes, they may become lethal toxin factories, already well adapted for survival inside the human body.

It is technically and economically unfeasible to extract militarily significant quantities of many potent toxins, such as shellfish toxin, from their natural sources. But prolific microorganisms fitted with toxin-producing genes can easily and cheaply mass-produce many of them. Further genetic manipulation could yield more efficient toxins—stable under a range of temperatures and resistant to degradation in the body.

- Safer experimentation. Improvements in physical and biological containment since the advent of rDNA technology have made the potentially grave dangers of BW experimentation far less daunting.
- Enhanced production efficiency. In the past a genuinely military BW production capability required massive, dangerous facilities and storage tanks that were difficult to conceal and maintain. New bioprocess technologies have drastically slashed the minimum size required for a BW production plant. And the time for the manufacturing process has been reduced by several thousandfold over earlier methods, according to the U.S. Army. They have eliminated the need for long-term maintenance of large, deadly stockpiles.
- Ethnic weapons. BW planners have dreamed for decades about precisely targetable weapons that would devastate the enemy but could never backfire or be used in retaliation. Since the advent of rDNA this fantasy has entered the realm of possibility. Specific ethnic or racial groups are susceptible to certain diseases or chemical poisoning, as a result of variations in natural resistance in the human gene pool. One example is valley fever, a fungal disease that has been closely studied by the U.S. Navy for decades. Certain studies suggest that blacks are far more susceptible to valley fever than whites. It may be possible to prey on such ethnic or racial groups by targeting a combination of these genetic factors.
- Biochemical weapons. The body produces tiny amounts of hormones and other substances that exert profound regulatory influence over moods, perceptions, organ function, temperature, and other essential physiological processes. The smallest imbalance can lead to severe illness, even death. Genetic engineering methods have made possible the manufacture of nearly unlimited quantities of these rapidly acting substances. This capacity has led to speculation about the weaponization of our own biochemical endowment.

- More potent chemical agents. The mode of action of nerve gas is being carefully studied for the development of antidotes. But elucidation of their mechanisms of action, combined with work in the chemical synthesis of toxins, is expected to lead to neurotoxic agents up to hundreds of times more potent than existing CW agents. Currently the superpowers' CW arsenals weigh in at more than 40,000 tons apiece. If this load could be cut to, say, 150 tons, it would be a phenomenal logistical windfall.

FANTASY VERSUS REALITY

The mere feasibility of these technologies does not make their implementation inevitable or logistically viable. Indeed, genetic engineering cannot "perfect" chemical and biological warfare, and the development of a genetically engineered arsenal would likely prove self-defeating—a destabilizing act with unpredictable and possibly dire consequences. But speculations on potential novel CBW applications are more than paranoia. The United States and possibly the Soviet Union are already traveling this dangerous road.

The 1980s are the decade of military biology. Nearly every item on the biotechnology wish list is a subject of intense U.S. military interest. For many years the U.S. budget for BW research stagnated at a modest level. But when the Reagan administration took control in fiscal year 1981, it initiated dramatic increases averaging 37.1 percent a year to $90.6 million in fiscal year 1985. This growth rate even vastly outstripped most other frenetically expanding Pentagon priorities.

The Department of Defense (DOD) claims that its entire BW budget and nearly all its research are defensive, unclassified, and open, as mandated by federal law. But the secrecy inherent in military operations makes it impossible to evaluate that claim with certainty. At the very least, official DOD figures grossly underestimate BW research funding. If we count all military spending on life sciences research—much of which presents clear applications to biological warfare—the yearly total exceeds $300 million. And some research that falls under the *CW* rubric—$235 million in 1984 alone—is clearly linked to *BW*.

As we will show in later chapters, the Reagan administration has also moved to constrict the flow of information on the nature of publicly acknowledged biological research despite its "unclassified" designation.

The DOD is working to establish an infrastructure of in-house biological expertise. In essence, it is "buying" a stable of scientists in academia and industry who increasingly rely on DOD contracts to weather cutbacks from other sources of federal biomedical research support or to survive the fiercely competitive climate within the biotechnology industry.

This major research effort relies on evaluation methods that would never pass muster in the NIH, which sets high standards for the review of biological research. Although much of the DOD program is described as "basic research," we will demonstrate that these methods obfuscate truly basic work and select for malleable scientists who are not positioned to understand the role their work plays in long-range military goals.

In this way the new biotechnologies are being applied to studies of dozens of putative BW agents—from organisms that cause Lassa fever, dengue, anthrax, and plague to the most lethal toxins. All of the "needs" of a biological warfare program are being pursued, including molecular genetics, rapid diagnosis, trials of protective gear, vaccine development and mass production, weaponized-agent survival tests, and the efficacy of insects as directed disease carriers.

The United States believes terrorists or poor, desperate nations may see a new generation of biological weapons as their path to power. CBW have been called the "poor man's atom bomb." Although no developing nation boasts the resources or expertise to mount a significant effort of this kind anytime soon, the fear may be well founded in the foreseeable future.

And U.S. claims that the Soviets have looked increasingly to biotechnology are certainly plausible; they could hardly have ignored military applications so obvious to the West. But no clear evidence shows that the Soviet Union has actually developed the full-blown gene wars program that the United States has suggested as justification of its own efforts. The Soviets are years behind the Americans in the biomedical sciences. Even if the Soviet will for military supremacy is as insatiable as the U.S. government suggests, the American lead is huge and growing.

Genetic engineering may never lead to the ideal germ weapon: unstoppable by its victims, yet controllable by its maker. But in military hands the ultimate life science already poses new and grave dangers. No supergerm is required to set off a chain reaction of suspicion and temptation that could unleash a new biochemical arms race.

The basis of U.S. BW research, according to the government, is defense against biological attack. But even the DOD acknowledges that in BW research the difference between offense and defense is purely a matter of intent. Moreover, this largely holds true for development, testing, production, and training.

Creating a truly effective weapon from an infectious agent requires intensive work to understand and master the microorganism; sophisticated manufacturing, test, and storage capabilities; delivery systems; the ability to protect one's own troops and population; effective methods of detection; and a reliable command and control apparatus.

To develop an acceptable biological warfare defense—though virtually impossible against a potentially infinite array of genetically altered BW agents—the same features are essential. The U.S. "defensive" program involves nearly all aspects of the BW process. It is not that offense and defense merely appear similar. They, in fact, share identical components.

A primary goal of this book is to put the U.S. military's intent and actions into scientific and historical perspective. At issue is the DOD track record on honesty, openness, concern for public health, and commitment to arms control.

We will show that the nation's historical record on CBW is replete with subterfuge, reckless experimentation, and rogue actions and is punctuated by violations of both domestic policy and international legal and moral norms. And the modern record is no more reassuring.

Of course, the United States is by no means a lonely outlaw. Molecular biology activities in the Soviet Union and other nations may deserve close attention. And the U.S. government is certainly not the world's most secretive. On the contrary, its relative openness made this book possible.

But the United States is the most powerful nation on earth, the leading nation in biotechnology. For four decades U.S. actions and policies have had the strongest influence on world behavior regarding CBW and all weapons of mass destruction. If modern biology is to be a tool for human benefit, not the seed of our destruction, then all its facets, including military applications, must be opened to new levels of public understanding and to careful public scrutiny.

2

INSTITUTIONALIZING SILENT DEATH: CBW DEVELOPMENT THROUGH 1969

THE DEMISE OF CAFFA

In 1346 the Black Sea port city of Caffa, ruled by Genoa, had been under siege by Tartars for three years. The Genoans had held firm within the massive city walls, confident that the amply stocked fortress would protect them indefinitely. The attacking Tartars were less sanguine. They had already lost many men to the hardships and battles of war. Bubonic plague—the Black Death that soon spawned the worst epidemic in history—had begun to depopulate Central Asian cities. The attacking soldiers were not exempt from the plague by virtue of military service.

"Infinite numbers of Tartars . . . suddenly fell dead of an inexplicable disease," wrote the Italian historian Gabriel de Mussis in 1348. "Arrows having been hurled from Heaven to oppress the pride of the Tartars . . . the humors coagulated in the groins, they developed a subsequent putrid fever and died, all council and aid of the doctors failing."

Then, de Mussis recounted, the attackers seized a desperate advantage. "The Tartars, fatigued by such a plague and pestiferous disease, stupefied and amazed, observing themselves dying without hope . . . ordered cadavers placed on their hurling machines and thrown into the city . . . so that by means of these intolerable passengers the defenders died widely. Thus there were projected mountains of dead, nor could the Christians hide or flee . . . they allowed the dead to be consigned to the waves. And soon all the air was infected and the water poisoned, corrupt and putrified."

Disease had established its place in history as a weapon of mass destruction. The Genoans gave up the city. Fleeing to Italy by sea, they hastened the plague's spread through Europe.

This is by no means the only or earliest case of warfare by contamination. The Romans fouled enemy wells with animal corpses 2,000 years ago—tactics copied during the American Civil War and the Boer War. British soldiers gave the American Indians "presents" that included blankets from a smallpox hospital. As many as 30 percent of the Indians died. Numerous survivors wore grisly scars as reminders of the treachery.

These episodes are somehow chilling beyond the usual horror of more "civilized," conventional warfare. The attackers turned public health in reverse and violated every moral precept and working principle medicine has aspired to since Hippocrates.

TECHNOLOGY TAKES COMMAND

Sophisticated chemical warfare preceded the use of biological agents. Review of the ancient Indian works *Ramayana* from 2000 B.C., *Mahabharata* from 1500 B.C., and *Artha-shastra* from about 400 B.C. reveals discussion of hypnotic, lethal, and tear gases as well as countermeasures, such as protective ointments and an antidote gas.

Similarly, in modern times, chemical war was waged before the introduction of biological weapons. On the afternoon of April 22, 1915, near the Belgian city of Ypres, Allied and German armies were dug into trenches a few hundred yards from each other. After days of incessant artillery fire the Germans opened 6,000 canisters, releasing billowing yellowish clouds of chlorine gas, which hugged the ground like fog and wafted over the unsuspecting French and Algerian lines.

The gas was so thick that the unprepared soldiers were blinded. Those not immediately incapacitated retreated wildly, choking to death. Next to the dying and wounded were puddles of yellow liquid coughed up from scorched lungs. French guns fell silent. A four-mile breach had been ripped in a front that had been impenetrable for months. The modern history of chemical warfare had begun.

The attack killed 5,000 men and wounded 10,000. Ironically, German commanders were nearly as shocked as the victims by the effects of gas. Never daring to suspect so decisive an impact, the Germans had failed to ready sufficient reserves to exploit their position. Instead of moving quickly to overtake the Allied positions, they simply observed the effects of their chemical attack and waited for the next opportunity.

In the next two years all the antagonists began mass production and use of chlorine, phosgene, which had similar effects, and finally, the most effective World War I chemical, the "king of gases"—dichloroethyl sulfide, known as mustard gas.

When first used by the Germans in 1917—coincidentally, again in Ypres—mustard gas was another surprise. Unlike the gas bombs that soldiers had learned to protect themselves against, mustard shells released oily brown liquid, giving off acrid fumes reminiscent of garlic. Initially the victims almost ignored the attack, seeing no dramatic ill effects. But within hours the Allied lines were a disaster scene. Contact with mustard vapor or liquid, from which several layers of clothing were no protection, produced blisters up to a foot long, blindness, choking agony, and sometimes death.

Increasingly effective combinations of agents and delivery systems were invented, only to be countered by improved protective gear. Researchers, for example, enhanced the efficacy of gas masks, which started as primitive flannel coverings soaked in a soda solution. In this way continual technological developments extended the gas war. This marriage of scientific and military minds has proved durable. It is a primary factor in chemical and biological warfare today.

By war's end 113,000 tons of poison gas had been used. Precise figures on the carnage are not available, but the war produced more than 1.3 million gas casualties, of whom about 91,000 died. The Germans, French, and British each suffered at least 200,000 casualties, and the Russians incurred double that number, despite an early withdrawal from the conflict following the Bolshevik Revolution. About 70,000 Americans, fully one-fourth of U.S. casualties, fell victim to chemical attacks. For all this misery, neither side was able to use gas to achieve other than a fleeting advantage.

In 1925 the Geneva Protocol outlawed use of chemical weapons but placed no limits on stockpiles. Most nations, including all world powers, signed the treaty. But many reserved the right to retaliate in kind, rendering the protocol a "no first use" pact.

With vast and growing budgets the U.S. Army Chemical Corps and equivalent units in other nations maintained their power and influence after the war. A CW infrastructure—combining the military bureaucracy and the scientific community that it commanded, hired, contracted, or wooed with appeals to patriotism—became firmly entrenched.

At the armistice in 1918 Edgewood Arsenal in Maryland may have been the largest single-purpose research and development facility in human history. Its cost: $40 million. Its goal: chemical weaponry. Within 79 research and administrative structures, about 1,200 scientists and technicians investigated thousands of poisonous substances as potential weapons. Edgewood's 218 manufacturing buildings could produce 200,000 chemical munitions per day. Discoveries at Edgewood and its counterparts in other countries, such as the Porton Down facility in Great Britain, made gas warfare increasingly sophisticated.

The pressure of increasing chemical armament weighed heavily on the Geneva Protocol. Finally, in 1936, it was first abrogated. Italy, which had signed and ratified the treaty in 1928, drenched Abyssinia (Ethiopia) with mustard gas. New technology provided the key motivation: Italy had just developed the first aerial gas-bombing techniques. The 500-pound bombs were timed to burst 200 feet above the ground, spraying a mustard mist in every direction.

This indiscriminate tactic proved devastating to the poorly equipped and trained Abyssinians. The spray killed 15,000 soldiers and many women and children. A greater casualty, however, may have been world confidence in the Geneva Protocol, as the great powers' attitudes toward CBW in the coming war would prove.

WORLD WAR II: CW COMES OF AGE

By the early years of World War II Europe stood poised for chemical Armageddon.

After a lull in CW production, which saw annual funding drop to about $1.5 million in the 1930s, U.S. research, development, and production (including BW and incendiary weapons) skyrocketed to $60 million in 1941. By 1942 this budget had hit a staggering $1 billion.

The British also maintained a major research effort, although they depended on the Americans to supply most of the actual gas as part of a long-term collaboration. The Soviets, debilitated by the Nazi invasion, nevertheless had a significant CW capability. The Germans had got a late start after their surrender in 1918. But by the mid-1930s they had renewed CW research, and by 1942 they had reestablished a CW production capacity of 12,000 tons per month.

All these countries tied their CW strategies to the air war. Increasingly sophisticated aerial spraying and bombing methods, combined with these monumental production levels, strongly suggested that CW would be the second war's decisive weapon.

Far more important, however, was a German discovery that changed the face of chemical warfare. The German chemical industry had long been the most sophisticated in the world. In 1937 the German chemist Gerhard Schrader stumbled upon a chemical four times as toxic as any war gas. The agent was tabun—the original nerve gas. A year later sarin—a related compound four times as toxic as tabun—was discovered. A third nerve agent, soman, even more potent than sarin, was developed by the war's end but never mass-produced.

These nerve gases operate by inactivating acetylcholinesterase, an enzyme that terminates the transmission of nerve impulses. By binding to acetylcholinesterase molecules, nerve gas creates wild, uncontrolled nerve signals and horrific death throes.

First the victim feels disoriented and breathing is labored. Soon foaming at the mouth and involuntary urination, defecation, and wild twitching of arms and legs take over as all bodily control is lost to violent convulsions. Death by suffocation follows; the diaphragm receives no signal to pump the lungs. This can take place in as little as a few minutes—faster if the victim breathes the agent, more slowly if it is absorbed through the pores of the skin.

Nerve gas was nothing short of a revolutionary weapon. The Germans harbored no illusions about its power; they produced secret stocks of more than 12,000 tons by 1945. It was the war's best-kept secret, safely held until the fall of the Third Reich. Amazingly the weapon was never used. Nor was any other CW used in Europe during the Second World War.

This was, in part, the product of Adolf Hitler's own experience during World War I. He was temporarily blinded by a massive mustard gas attack during the final Allied offensive. Particularly in the early stages of the war Hitler was not ready to provoke what was sure to escalate into an uncontrollable eruption of gas. The Nazis also realized that gas would bog down their heady invasion pace by requiring soldiers to don bulky protective gear.

In the late stages of the war German commanders greatly feared retaliation in kind against their civilian population, particularly after the unprecedented destruction of the Allied saturation firebombings. They also believed incorrectly that their enemies, too, had discovered nerve agents. In fact, the Allies were shocked when they fell upon German nerve gas stocks in 1945.

Hitler might have been persuaded by the end of the war to unleash nerve gas in desperation, but by then his bomber force, decimated by Allied raids and antiaircraft, was insufficient to deliver the weapon in decisive quantities.

Like the Germans' reticence, the Allies' failure to use chemicals had little to do with lack of planning or the tether of international law. In 1944, after Germany's V-1 rockets had wreaked havoc on British cities, Winston Churchill issued an order to his generals that shows how close the world came to the poison gas debacle so many expected. Churchill wrote in a top secret memo:

> It is absurd to consider morality on this topic when everybody used it in the last war without a word of complaint from the moralists or the Church. On the other hand, in the last war the bombing of open cities was regarded as forbidden. Now everybody does it as a matter of course. It is simply a question of fashion changing as she does between long and short skirts for women. . . .
>
> We could drench the cities of the Ruhr and many other cities in Germany in such a way that most of the population would be requiring constant medical attention. . . . I want the matter studied in cold blood by sensible people and not by the particular set of psalm-singing uniformed defeatists which one runs across now here now there.

Cooler heads prevailed. Churchill's advisers argued that beyond the expected retaliation, gas might actually prolong the war. Planes needed to bomb the Nazi industrial base would have to be diverted for attacks on civilian populations.

Saul Hormats, a high official in the U.S. Army Chemical Corps during the war, wrote recently that like the Nazis and British, many U.S. tacticians believed "that introducing CW into battle [in Europe] would not change the outcome but would only prolong it [and by slowing troop movement] result in more casualties to no military purpose." This conclusion showed how little the ultimate utility of poison gas had been improved by technological developments between the world wars.

His personal revulsion to gas spurred President Franklin Roosevelt to overrule repeated attempts on the part of American military chiefs to use it against the Japanese. Pressure was particularly keen during the late stages of the Pacific war—to reduce the massive casualties suffered at such islands as Iwo Jima. Roosevelt's stand was a lonely moral victory. In general, it was the amalgam of tactical concerns, intelligence bungling, and fear that held CW in check during World War II.

BW "PIONEERS"

The biological sciences were advancing rapidly by World War I. But fighting disease was still a chancy proposition. Antibiotics, the greatest weapon in the battle against bacterial infection, were not discovered until the early 1940s. Before then a primary deterrent to the use of disease as a weapon of war was still firmly in place: Attackers could not protect their own troops and populations. In isolated cases German spies were accused of biological sabotage during World War I, but no nation maintained a significant BW research and development program.

Still, when the horror of gas warfare led to the Geneva Protocol, its authors, recognizing the links between chemical and biological methods, included a precautionary prohibition against the use of "bacteriological methods of warfare."

Not long afterward the biological arms race began in earnest. This time Japan played the leading role. Army Major Shiro Ishii apparently found the Geneva mention of bacteriology irresistibly suggestive and convinced his superiors to establish a germ warfare research center in 1935. The effort gained momentum soon after, when the Japanese arrested alleged Russian spies and charged them with carrying the bacteria that cause cholera—ostensibly to sabotage the Japanese water supply. Ishii's program complemented Japan's major CW research and development effort.

By 1939 Ishii, then a general, had established the first BW research and production facility, called Pingfan, near the city of Harbin in occupied northeastern China. It was followed by eighteen smaller installations.

The Japanese designed a wide range of weapons. One of the most successful efforts was the weaponization of the organism that causes anthrax, a highly infectious disease of cattle and sheep. Shrapnel wounds from an anthrax bomb cause grotesque black skin lesions, leading to blood poisoning and death in nine of ten victims. If anthrax bacteria are inhaled, the disease is nearly always fatal. Because the organism forms hardy spores that can survive indefinitely, even in harsh environments, it clearly has the robust quality necessary for an effective BW agent. Pingfan also produced weapons from many other deadly diseases, including cholera, typhoid, and dysentery.

One shocking element of Japan's program sets it apart from nearly all BW research before or since: human guinea pigs. At least 3,000 Chinese, Korean, Soviet, American, British, and Australian prisoners of war died horrific deaths at the hands of Pingfan technicians.

The work included trials of anthrax and gas gangrene bombs. Prisoners were tied to stakes, their buttocks exposed to the shrapnel flying from a bomb detonated by remote control. The course of the disease was meticulously tracked and recorded as the victims died in agony. Other prisoners were infected with organisms causing cholera and plague, only to be dissected—sometimes while still alive—to monitor the progressive degeneration of their internal organs. Chinese women "were infected with syphilis with the object of investigating preventive means against this disease," according to a 1949 Soviet war crimes trial record.

The Soviets estimated that at its peak Pingfan was an industrial-scale plant capable of producing more than 500 pounds of bacteria per day and breeding 500 million plague-carrying fleas a year.

By 1939, during its China invasion, Japanese commanders felt ready to put to use their years of CBW work, beginning with mustard gas attacks. BW soon followed. Several sources, including the People's Republic of China official archives, describe BW attacks on at least eleven Chinese cities through 1944.

In one case low-flying Japanese planes dropped grains of wheat and rice and pieces of cotton and paper over the Hunan Province city of Changteh in November 1941. After several townspeople had died from a disease resembling plague, autopsies revealed the presence of the plague bacterium, *Pasteurella pestis*. The Chinese estimate they suffered 700 deaths from plague attacks alone. The Japanese are also thought to have poisoned more than 1,000 Manchurian wells with cholera, typhoid, and dysentery.

In the face of a mountain of evidence Japan continued to deny the existence of the germ warfare unit until 1982—and it still officially denies the human experimentation atrocities.

THE RISE OF BW

Military planners from other nations also began to see BW as a useful indiscriminate weapon by the 1930s. Anything so hard to defend against offered strong offensive incentives.

A wide range of agents was seen as having the potential for everything from sabotage to mass bombings. BW was relatively inexpensive compared with the manufacture of other munitions. And these weapons represented a potent psychological tool: Odorless, invisible, and deadly agents were expected to provoke such terror as to destroy troop morale and induce panic among civilians.

During this period tacticians also recognized that because of unique property—self-replication—BW agents had major advantages over other weapons of mass destruction. With the most toxic chemicals and powerful explosives of the period, quantities in the tons were needed to overwhelm even a small area. Whole squadrons of bombers might be required to incapacitate a major industrial district or city.

Tiny quantities of BW, however, could theoretically do the same job. As an added attraction, the physical environment would be preserved for occupation. Experts estimate, for example, that effectively disseminated, as little as eleven pounds of plague bacterial paste could have a direct military impact on one square kilometer—and could promote the spread of disease over wider areas.

Despite the German discovery of nerve gas, and macabre experiments conducted on concentration camp prisoners, including covering them with typhus-infected lice, Nazi BW research lagged behind that of other nations. The Nazis never established a workable, systematic program for biological warfare.

Knowledge of the Soviet Union's efforts with BW during the war came largely from German intelligence files captured in 1945. From these documents the Western powers concluded that the Soviets had established a BW capability in the mid-1930s. In 1938 a Soviet army general confirmed that at least some BW work was taking place. His country adhered to the 1925 ban on the first use of BW, he said, "but if our enemies use such methods against us, I can tell you that we are prepared—fully prepared—to use them also, and to use them against aggressors on their own soil."

Unverified accounts from Soviet defectors and U.S. intelligence files claim that anthrax and cholera were studied for their potential as weapons and that at least two plague weapons were developed. One of these involved infecting a particularly aggressive type of gray rat and dropping the infected rodents from low-flying planes.

The British began BW research in earnest in 1934. Like the Japanese, they were stimulated to take germ warfare seriously by the Geneva Protocol. At first the work was defensive. By the late 1930s stocks of vaccines, fungicides, and insecticides had been procured to counter any Nazi attack. By 1940 German victories had spurred the British to weaponize biological agents.

A secret laboratory was set up at Porton Down. By the war's end its work had grown to mammoth proportions. The British produced *5 million* crude anthrax weapons and collaborated with the Americans to develop other biological munitions. Despite overwelming evidence from intelligence leaks and declassified documents, the British government steadfastly holds to this day that it has never acquired biological agents in quantities large enough for offensive use.

The major American BW effort began in 1941. Of the multibillion-dollar CBW budget during the war years, more than $40 million—a tidy sum even by today's standards—was spent on BW plant and equipment purchases alone. Weapons research, development, and testing were conducted by a phalanx of 4,000 scientists and technicians.

Camp Detrick (renamed *Fort* Detrick after the war, reflecting biology's secure niche in U.S. strategy), in Frederick, Maryland, was established as the chief research installation. Detrick scientists carefully examined a Pandora's box of diseases—from the common cold to anthrax, typhus, yellow fever, plague, botulism (commonly known as a deadly food poison), and dozens of others. They also conducted large-scale antianimal and anticrop research.

In 1943 the Americans manufactured the Allies' first biological bomb, a four-pound anthrax device based on a British design. Prototypes were tested at the Dugway Proving Ground, a desert range larger than the state of Rhode Island, near Salt Lake City, Utah.

The American industrial base, untouched by the war, was the workhorse of BW development for the Allies. In the United States BW systems, including aerial sprays, antipersonnel mines, and a variety of bombs, were perfected and manufactured.

The Vigo, Indiana, production plant was the effort's crowning achievement. It boasted 20,000-gallon tanks that could produce 500,000 anthrax bombs a month. It was capable of growing vast quantities of botulinus toxin for botulism bombs. In this form the toxin remained potent and usable for months or longer. Ounce for ounce, it is one of the most deadly substances known, hundreds of times as potent as nerve gas. Because the end of the war nearly eclipsed the plant's construction, it was never used, although Vigo stood as an unambiguous testament to the importance of BW in military thinking by 1945.

Despite such awesome preparations by many nations, BW rarely moved beond the test range. With the exception of Japanese attacks on China, antipersonnel BW was never used on a mass scale.

The Polish underground is thought to have killed several hundred Nazis by covertly infecting them with typhoid fever. And there has been some speculation that the United States may have made biological attacks on Japanese and German crops. But in the final analysis, BW traveled the path of its chemical counterparts. A combination of tactical complications, fear of retaliation, and timing considerations spared most of humanity the trauma of biological warfare during World War II.

POSTWAR STRATEGY

Wartime restraint was followed by a confused, contradictory CBW posture reflecting profound differences within U.S. military and civilian policy-making bodies. These conflicts indicated a deep chasm between the acknowledged and actual objectives.

U.S. policy development was predicated on the nation's formidable CBW capability at war's end. The American CW inventory, the world's largest, stood at 135,000 tons—more than all the gas used by all sides during World War I. The supply was eighteen times Japan's arsenal at the time of surrender. Although some stocks of World War I-era chemicals were destroyed in 1945, they were quickly replaced by far more potent nerve agents.

The postwar U.S. lead in chemical readiness, however, was attributable to more than just its gigantic supply of CW agents. American superiority was fueled by a willingness to try out any new chemical, no matter how unlikely. Even unsuccessful efforts ruled out unprofitable lines of inquiry and provided guidance for defensive countermeasures.

One project, for example, code-named Who, Me?, was designed "to develop a material with a generally repulsive and persistent odor suggestive of personal uncleanliness," noted a 1944 document released decades later. "The ultimate in repulsive odors," the summary aptly explained, "is one that stinks 'to high heaven.'"

In a more sinister vein, Robert Oppenheimer, leader of the atomic bomb design team, was involved in the development of radioactive poison in 1943. "I think that we should not attempt a plan unless we can poison food sufficient to kill a half a million men, since there is no doubt that the actual number affected will, because of non-uniform distribution, be much smaller than this," Oppenheimer coldly suggested in a recently declassified letter to fellow atomic pioneer Enrico Fermi.

Organizational drive and momentum also led U.S. biological research and development (R&D) forward ever more quickly, even after it became clear late in the war that the Axis powers were in no position to mount a significant BW attack.

In 1945 the U.S. government solicited what was to become a profoundly influential report from a leader of the wartime germ warfare effort, George W. Merck, head of the Merck pharmaceuticals company. He recommended that full-scale research and development be continued. Merck's recommendations were largely accepted, although production capabilities were reduced. The mammoth Vigo plant was eventually turned over to a drug manufacturer, Pfizer Inc, with the contingency that it could easily be reconverted to BW production. Camp Detrick was chosen as the chief site for ongoing research.

After the war the battered European allies were happy to defer the major CBW responsibility to the United States while they rebuilt the shambles of their cities and economies. Wartime cooperation continued, however. Detrick and England's Porton Down collaborated on weaponizing and defending against chemical and biological warfare. An agreement was also set up with Canada and Australia, whereby research was conducted by Porton Down, development by Detrick, and the others' vast, little-used territories became field test sites.

As the Red Army thundered through eastern Germany at the war's climax, some nerve gas, a manufacturing plant, and engineering plans fell into its hands. The plant was removed, piece by piece, back to Russia. The Western powers also learned the Nazi secrets but didn't have the good fortune to capture a completed facility. As the cold war took shape, this Soviet capability weighed heavily on military minds. More important, the "Soviet threat" replaced wartime enemies as the ostensible reason for continuing the U.S. CBW program, a rationale that has survived intact to this day.

The Soviets apparently built a massive chemical arsenal. But forming reliable estimates of the size and scope of the Soviet program—vital to effective policy and countermeasures—was (and has always been) a vexing problem for the West. Intelligence figures on Soviet CW stocks varied widely in the years following the war.

It was also assumed that the Soviets plowed ahead with biological warfare research and development. But, again, estimates of the Soviet *BW* research and products—some of which suggested vast superiority over the United States—spanned a broad range. This suggested a reliance on the same kind of highly speculative intelligence data used to project Soviet *CW* capability. In any case, independent scientific authorities agreed that whatever the Soviet Union's goals may have been, its command of the biological sciences lagged far behind that in the United States.

THE JAPANESE DATA GAP

The Soviets may have had more reason to accelerate BW research after the war than the Americans: They soon learned (though the American public remained ignorant of it) that the unique Japanese program of human research had fallen intact into American hands.

Before the Soviets invaded Japanese-held Manchuria a week before Japan's surrender, Pingfan closed up shop. Remaining subjects were killed and burned. BW agents and vectors—living organisms, such as insects and animals, that spread disease—were incinerated. Buildings and equipment were destroyed or disassembled. Experimental data and laboratory slides were sent secretly back to Japan.

Recently declassified documents tell a remarkable tale about the breadth of the American "national security" mentality: In 1947 U.S. occupiers captured some leaders of the Japanese BW program. In top secret cables army officials told Washington about their prisoners and made a policy suggestion: "[Lieutenant General Ishii, director of BW work] states that if guaranteed immunity from 'war crimes' in documentary form for himself, superiors and subordinates, he can describe program in detail. . . . [A]ggressive prosecution will adversely affect U.S. interests."

Reportedly with the approval of General Douglas MacArthur, the commander of the occupying force, the suggestion was readily accepted. Another top secret document concluded:

> Because of the vital importance of the Japanese BW information . . . the Japanese BW group should be informed that this

> Government would retain in intelligence channels all information given by the group on the subject of BW. This decision was made with full consideration of and in spite of the following: That its practical effect is that this Government will not prosecute any members of the Japanese BW group for War Crimes of a BW nature. . . . That there is a remote possibility . . . that American prisoners of war were used for experimental purposes by the Japanese BW group.

(The last consideration was substantiated beyond any doubt by documents obtained by U.S. journalist John Powell in the early 1980s. The declassified U.S. reports show that the military had learned during the occupation about atrocities against American POWs.)

The blanket immunity allowed thousands of Japanese BW researchers to slip unnoticed back to the civilized world. Some ranking officers went on to lead prestigious Japanese microbiology or medical institutions. Hisato Yoshimura, who had directed experiments involving the freezing and thawing of living victims' limbs, earned large fees as a "freezing consultant" for a commercial fishery. He also became the first president of the Japan Meteorological Society. Yoshimura was eventually forced to resign from the society after his wartime activities were revealed in the early 1980s. But many of his cohorts lived long, undisturbed, and esteemed professional lives.

It is difficult to overestimate the significance of their legacy. By 1945 the American effort in BW was already impressive. But with the Japanese data it gained the only known scientific studies of the effects of germ warfare on human beings. Much of the data, "human pathological remains," and descriptive information from the tests—both human experiments and "field tests" of BW against the Chinese—were transferred intact into American hands.

Dr. Edwin V. Hill, then chief of the basic sciences division at Camp Detrick, indicated the data's extraordinary military value. "Evidence gathered . . . has greatly supplemented and amplified previous aspects of this field," he wrote in 1947. "Such information could not be obtained in our own laboratories because of scruples attached to human experimentation. . . . Furthermore, the pathological material which has been collected constitutes the only material evidence of the nature of these experiments."

The Soviet war crimes trial at Khabarovsk in 1949 gave the world the first detailed look at the Japanese BW program—through the eyes of victims as well as low-level technicians captured in 1945. The Soviets publicly accused the United States of shielding the masterminds of Japanese germ warfare in order to use their information for BW preparations. The United States brushed off the charge as "propaganda." Soviet motives aside, however, the accusation was entirely accurate.

And as Hill pointed out, the Japanese data were unique. The Americans learned from Yoshimura and others that the Soviets did not capture militarily significant information when they overran Pingfan in 1945. A U.S. war crimes trial akin to Khabarovsk was rejected because it would have revealed the data to the world.

OPERATION BLUE SKIES

After the war U.S. military circles considered BW the ideal strategic complement to atomic weapons. Biological warfare was seen as a major alternative in minor wars or conflicts in which atomic bombs would be difficult to use. This logic was particularly fashionable at the time because atomic weapons technology had not yet reached the mass production stage.

But nuclear weapons gradually grew to dominate overall defense strategy. As atomic advocates gained ascendency at the Pentagon, U.S. posture on both biological and chemical arms began to change. By the late 1940s powerful military leaders had concluded that CBW were troublesome anachronisms. With the ultimate strategic weapon in hand, they concluded, it didn't make sense to continue concentrating on socially repugnant, controversial weapons of questionable reliability.

The CBW program did not vanish entirely, however. The massive CBW bureaucracy still wielded considerable influence. Those in the government who discounted biological warfare as a strategic threat had little or no knowledge of the Japanese data. In contrast, Chemical Corps leaders were fully cognizant of the extent and implications of the capture of the Japanese BW archives, which fueled their beliefs that BW was a plausible threat and that the United States held a substantial edge on the competition. They remained convinced that germs and chemicals could be used effectively for sabotage as well as for other tactical and strategic missions.

But for a time their work was subjugated to other military and civilian priorities. By 1950 the bloated wartime CBW research and development budget had plummeted to less than $7 million. It averaged a mere $19 million per year in the period from 1946 to 1957, and the very survival of the corps seemed questionable.

But in 1959 corps leaders decided to take action against the public squeamishness and bureaucratic timidity that were strangling their efforts. They initiated Operation Blue Skies. This brilliant publicity maneuver, conceived by outside public relations consultants, was designed to teach people to "love that gas," as one analyst quipped. Numerous articles placed in the popular press, congressional appearances, lectures, speeches, and symposia all recycled the antiquated notion that chemical and biological weapons—sometimes less lethal than bombs or bullets—were the most "humane" of all weapons. One article went so far as to suggest that CBW research and development offered the promise of "war without death."

Simultaneously Chemical Corps representatives, who argued forcefully that CBW disarmament presented insurmountable verification problems, began to release precise estimates of Soviet CBW strength for the first time. In 1960 the head of army research told a congressional committee that one-sixth of the Soviet arsenal consisted of chemical munitions. If true, this massive buildup was a frightful problem indeed. It would warrant a major U.S. response.

According to a "normally reliable" source of Pulitzer Prizewinning reporter Seymour Hersh, however, the calculations were of dubious merit. "The Army computed the roof size of the Russian sheds," the source noted, "figured out how many gallons of nerve gas could be stored in a comparably-sized shed in Utah, added a 20 percent 'fudge' factor, and came up with the estimate."

It is uncertain whether the Chemical Corps really took seriously its "humane weapons" fiction or inventive intelligence reports. But Operation Blue Skies worked. By the beginning of the Kennedy administration in 1961 there were considerable rumblings about excessive reliance on nuclear weapons at the expense of flexibility. By then CBW lobbyists had filled the heads of high officials and the public with fantasies of benign weapons that destroyed the will to resist without killing the enemy or harming valuable property. This was comforting imagery compared with nuclear incineration.

The result: CBW research and development appropriations soared. By 1964 they had reached $158 million, not including $137 million for procurement and vast sums used to build research, development, and production facilities.

The money bought a vast CBW effort. Even during the funding

doldrums of the late 1940s to late 1950s the army had produced and tested hundreds of drugs, chemicals, and biological agents. By the early 1960s the DOD was able to indulge its most outlandish impulses and build the grandest empires—almost all in secret. Even now the program's full extent is unknown, although parts of the puzzle can be pieced together from declassifications and leaks over the years.

We know, for instance, that at least 78 leading universities and 161 private companies accepted fat contracts to pursue open or secret CBW research and development projects that ran the gamut from molecular biology to munitions. In some cases unwitting academic scientists were recruited to conduct BW research; a few of them surmised the true goals of the epidemiology and field ecology studies and suspended the work angrily, to the chagrin of their institutions.

And we know that the army also conducted major studies of animal and insect vectors. Fort Detrick developed controlled conditions for breeding *Aëdes aegypti* mosquitoes—which can transmit yellow fever—by feeding them a mixture of sugar syrup and blood. "Fort Detrick's laboratories were capable of producing half a million mosquitoes a month, and the Engineering Command designed a plant capable of producing 130 million mosquitoes a month," according to a declassified army report. Many of these mosquitoes were infected with yellow fever virus, which they successfully transmitted to lab mice. Hundreds of thousands of uninfected *Aëdes aegypti* were dropped from planes over U.S. cities to map the range of their spread. Detrick labs also produced plague-infected fleas, ticks with tularemia, and flies carrying cholera, anthrax, and dysentery.

Beyond animal testing of agents, the army wanted human data to augment what had been obtained from the Japanese. Operation Whitecoat provided the answer.

At least 2,200 patriotic volunteers from the Seventh-Day Adventist Church sampled many of Detrick's wares from the 1950s through the early 1970s. Church members believe strongly in the Ten Commandments, which preclude their engaging in actual combat. A church leader described their rejection of soldiering, while participating in BW testing, as evidence that they thought of themselves as "conscientious cooperators" rather than "objectors." The Adventists provided a steady source of human test fodder until Whitecoat was terminated in 1973. The army claims few of these volunteers suffered ill effects from their service. But as will be discussed in Chapter 5, this claim is highly suspect.

The Stockholm International Peace Research Institute (SIPRI), a leading arms control think tank, conservatively estimates that by 1969 the United States had stockpiled 40,000 liters of antipersonnel BW agents and at least 45,000 toxin-containing bullets and shrapnel bombs. Also on hand were more than five tons of antiplant BW agents.

GOING FOR VOLUME

In 1969 Matthew Meselson, a Harvard biochemist and one of the nation's leading CBW authorities, told the Senate Foreign Relations Committee that "the field testing of live biological weapons, and especially the outbreak of actual biological warfare, would constitute a menace to the entire human species."

Unbeknownst to Meselson at the time was the military's live-pathogen field test program. From 1951 through 1969 hundreds, perhaps thousands of open-air tests of organisms causing disease in humans, animals, and plants were conducted. The army used Utah's vast Dugway Proving Ground and more than two dozen other sites, including, incredibly, unrestricted public lands. The trials involved virtually everything in the arsenal from wheat stem rust and rice blast to anthrax and plague.

Public knowledge of these tests comes largely from data released during U.S. Senate hearings in 1977. Despite government claims to the contrary, these data are far from complete, and even official documents show many tests taking place over a period of months, with no indication of how extensive and frequent each series may have been.

At Dugway the army apparently infected and released many species of animals. University of Utah researchers then conducted periodic assays to monitor the rate and extent to which the disease spread through the animal population. Despite army denials, university contract reports indicate that infected insects may have been released as well.

Often biological and chemical bombs were dropped from airplanes. Cognizant of the taxpayers' trust, the army also dumped or fired chemical or biological munitions from 50- to 325-foot towers, to save on airplane fuel.

A major testing program only barely mentioned in the 1977 hearings involved millions of birds in the South Pacific. In conjunction with the prestigious Smithsonian Institution, the army banded these "avian vectors" to check their flight patterns. Substantial evidence suggests that some birds were infected with aerosolized Q fever or Venezuelan equine encephalomyelitis (VEE), a highly infectious disease causing acute flu symptoms in humans. The researchers initially assumed that the birds would stay within a 4-million-square-mile grid of the Pacific, but experts now believe that some reached continental landmasses.

Even more extensive research was conducted using biological simulants—ostensibly innocuous agents that mimic the behavior of pathogens. Such agents can cause illness in children, the sick, and the elderly. In the late 1970s public outrage was stirred by revelations that the army had secretly sprayed simulants from bogus suitcases in major airports and bus stations to check how far and fast they could be spread.

In another episode an army operative broke a simulant-filled light bulb in a New York subway station. Various points along the subway system were then checked for evidence of the organism. The simulant *Serratia marcescens* was sprayed in massive quantities over San Francisco Bay and is believed to have caused a minor epidemic, killing at least one elderly man.

The stupefying volume of army testing is a story in itself. The Dugway Proving Ground library alone now holds about 67,000 reports averaging 150 pages in length, according to Colonel R. Rex Brookshire II, chief counsel of the Army Test and Evaluation Command. As many as 10,000 of these documents involve field tests of live pathogens.

This remarkable arsenal of U.S. data, when combined with Japanese human trial results, constitutes the blueprints on toxicity, animal epidemics, munitions, agent aerosolization and concentration, and climatic effects that are essential to the use of BW, regardless of any changes or "improvements" that modern biotechnology may impose on the organisms themselves.

STRIKING FIRST . . . OR NOT?

"We shall in no circumstances resort to the use of such weapons unless the first use of them is by our enemies." So began a 1943 pledge by President Roosevelt that constituted official U.S. policy on CW. Because CW and BW are closely linked in military policy and planning, the operative assumption was that the United States also rejected first use of BW.

A credible evaluation of current U.S. motives must appraise the reliability of stated or implied policies. Substantial evidence suggests that the Pentagon never adopted Roosevelt's chemical pledge. Moreover, there are strong indications that the United States has never ruled out a biological first strike.

This *secret* policy leaked into the army's open literature in the 1950s. From before World War I until 1955 the principal army manual discussing CBW described it as only retaliatory. But the 1956 version of this manual contained a statement indicating merely that no *treaty* restricts U.S. use of CBW. And in 1958 Major General William M. Creasy, chief of the Chemical Corps, acknowledged to a congressional committee that official CBW policy had changed in 1956. But the new policy itself was deleted from the heavily censored hearing transcript.

In 1959, when a resolution reaffirming the *stated* no first use policy was introduced into the House of Representatives, the Defense and State departments opposed it vigorously. "As research continues, there is increasing evidence that some forms of these weapons, differing from previous forms, could be effectively used for defensive purposes with minimum collateral consequences," the DOD argued. This telling admission strongly suggests a policy permitting CBW response to enemy attack—regardless of whether the enemy had used them first. The House resolution was defeated.

The controversy then receded from public view but continued to simmer behind the scenes until the mid-1960s, when U.S. use of tear gas and chemical herbicides in Vietnam stimulated considerable international outrage. White House adviser Donald F. Hornig raised BW policy in a memo to President Lyndon Johnson.

"In explaining the use of riot control agents and defoliants in Vietnam, senior officers of your administration have made it clear that it is against our policy to initiate the use of chemical warfare," Hornig said, reflecting the U.S. view that such agents are not chemical weapons. "There has not, however, been a comparable public statement concerning a policy of 'no first use' of biological weapons," he continued. "In the absence of a publicly stated position, this leaves us particularly vulnerable to charges that it may be our intention to employ such agents." Rhetorical niceties aside, Hornig was acknowledging that the operative U.S. policy did not exclude first use of BW.

Hornig went on to say that the president's Science Advisory Committee "believes it extremely unlikely that we would, in fact, consider initiating the use of these [biological] weapons in a military conflict" and that the committee recommended a no first use policy statement. Johnson never made such a pledge.

Amid the often complicated and obscure arguments, a profound yet simple point somehow eluded the debaters until years later: Biological weapons are inherently useless for tactical retaliation and inferior to nuclear weapons for strategic retaliation. Because biological weapons are slower-acting and harder to control than other weapons, any nation that would launch a BW attack is likely to have taken precautions to protect its troops. Therefore, retaliation in kind primarily would affect civilians. Ironically, the only logically effective BW policy is a first use policy.

The most important test of policy is practice. No government will voluntarily confirm that it has abrogated international standards and stated policy, but substantial evidence indicates repeated U.S. first strikes through 1969. (The most recent incidents are discussed in Chapter 3. Modern U.S. war-fighting strategy is covered in Chapter 6.)

Declassified documents show that the Office of Strategic Services (OSS, the CIA's World War II-era predecessor) planned and may have conducted biological and chemical attacks in the 1940s. And the military may have used an anticrop weapon—Colorado beetles—to besiege Germany in 1944. The beetle problem was so severe that Gerhard Schrader, inventor of nerve gas, was diverted from weapons research to come up with an effective pesticide to save the potato crop. And in Japan a major rice blight followed U.S. bombing runs in 1945.

More compelling evidence is available regarding alleged attacks during the Korean War. In February 1952 the North Koreans and their Chinese allies paraded captured U.S. bomber pilots who admitted dropping "germ bombs." The United States quickly labeled their statements "fabrications," claiming that the pilots had been brainwashed. The Chinese responded by forming an international investigative commission boasting experts from the Soviet Union, Italy, France, Sweden, Brazil, and Great Britain. Its 700-page report, issued in October 1952, concluded that "the peoples of Korea and China did actually serve as targets for bacteriological weapons." The report indicated the Americans experimented with everything from fountain pens with infected ink to a range of vector techniques, including anthrax-tainted feathers and plague-ridden fleas and lice.

The United States strenuously denied all charges. When the Chinese suggested that the methods were reminiscent of attacks by Japan against China, the Americans—who held the Japanese data—said they knew of no such atrocities. The United States urged that the United Nations (UN) conduct a second study, but this was blocked by the Chinese and North Koreans.

True or not, the allegations damaged U.S. credibility. The embarrassment was so acute that the State Department immediately launched a detailed, worldwide analysis of the "communist propaganda campaign." (Analyses to assess the need for CBW damage control by no means ended with the Korea case. Following disclosures in 1976 of BW simulant tests in U.S. cities, the army conducted a detailed survey of news reports.)

"The Kremlin has succeeded in making 75% of the people on both sides of the Iron Curtain *aware* [emphasis in original] of germ warfare—and aware of it in terms of a relationship to the United States," President Eisenhower was warned by adviser C. D. Jackson in April 1953. Jackson advocated a major propaganda counteroffensive.

THE ESPIONAGE CONNECTION

In 1975 Senate hearings revealed the CIA as a major player in chemical and biological warfare. Its work was conducted under a program code-named MKULTRA. By the CIA's definition, MKULTRA was "concerned with the research and development of chemical, biological, and radiological materials capable of employment in clandestine operations to control human behavior." Its trademark was the routine use of unwitting subjects, researchers, and institutions.

With 149 subprojects from 1953 until the late sixties MKULTRA (and its successor programs, which continued into the early seventies) were a spymaster's dream. The CIA encouraged its agents to think creatively and provided the funds to turn fantasy into reality. MKULTRA projects were far more secrective than even the normal CIA bill of fare. And the handful of top officials who knew about MKULTRA maintained a hands-off policy.

This latitude and lax supervision contributed to the pursuit and failure of many farfetched notions, such as intensive research into parapsychology. The CIA was interested in whether mental telepathy could be used to control people in distant locations. The agency also explored the use of clairvoyants to predict the future. MKULTRA's freewheeling nature pushed it beyond mind control per se into CBW research and development—and possibly covert biological warfare.

Subproject 146, for example, devised in 1963 and 1964, involved an internationally known plant pathologist "who will assist in developing a philosophy of limited anticrops warfare," according to a heavily censored record. He was used "to formulate a basic approach to an attack of [deleted]," it noted. The target was blacked out, but the document specifically mentions sugar as the crop of interest. Cuba, which has made several accusations of CIA biological weapons attacks aimed at destabilizing its economy, claims its poor sugar crop in 1964 was the first such target.

This work was closely linked to the army—including direct agency access to Fort Detrick—from 1952 on. The agency paid Detrick $100,000 per year in the early sixties for a supply of BW agents and delivery systems. The army and the CIA pooled their resources on a variety of schemes to assassinate Cuban leader Fidel Castro, according to a declassified CIA document. Poisoned cigars and pills to be dropped into milk shakes—among the Cuban's favorite foods—were two such ideas.

But the strongest MKULTRA emphasis involved psychochemicals. From 1954 to 1963 the CIA tested LSD on unwitting subjects picked up in New York and San Francisco bars by agency-hired prostitutes. The subjects were lured to "safe houses"—essentially CIA bordellos—where the drug was slipped into their drinks. A voyeuristic agent watched and took notes from behind a one-way mirror.

The army, too, was enamored of LSD as well as quinuclidinyl benzilate, code-named BZ, a far more powerful hallucinogen, which can incapacitate a person for twenty-four hours. The two drugs were Operation Blue Skies' chief discoveries. Working with the CIA, the army gave LSD to thousands of volunteer soldiers in the 1950s and 1960s. They were interrogated, drove tanks, and operated radar while under the influence.

Other subjects were unwitting. One of these, Fort Detrick BW researcher Frank Olson, committed suicide by crashing through the window of a tenth-floor office after his LSD trip had reduced him to a psychotic state. The resulting press attention forced LSD testing into the open. The army ultimately decided that LSD was too intoxicating for effective interrogation or manipulation.

BZ was considered promising by the army for a time. But it was also abandoned as too powerful and unpredictable. Always thinking on a grand scale, the army manufactured tons of BZ, which still await disposal. During the 1950s Edgewood Arsenal each month reviewed 400 chemicals rejected by pharmaceutical firms because of negative side effects. But it never found a drug with the potential of the two hallucinogens.

It is hardly surprising that the army and CIA would seek the advice of civilian experts. In wartime, industry and academic scientists had always provided the intellectual capital that produced major CBW developments. The most prominent example is the discovery of nerve gas by the German chemical industry.

Under MKULTRA, the CIA used at least forty-four institutions of higher learning, twelve medical facilities, and three prisons. At times institutional directors and individual researchers were cognizant of the funding source and its goal. More often, however, they had no idea about CIA sponsorship or intentions and would have withdrawn if they had. And none of the subjects, of course, knew of the CIA sponsorship. The minds of the criminally insane were a favorite proving ground. Prisoners in the Vacaville, California, Medical Facility were probed thoroughly after imbibing heroin, LSD, and a variety of other drugs.

"Research in the manipulation of human behavior is considered by many authorities in medicine and related fields to be professionally unethical," placing researchers' reputations in jeopardy, the CIA inspector general aptly stated in a 1963 report, thus justifying MKULTRA's extreme secrecy and procedures. The full extent of the program will never be known since many agency records were illegally destroyed in the early 1970s.

By the same token, the BW attacks on Germany, Japan, and North Korea will probably never be decisively verified or disproved. Throughout its history the U.S. CBW program has worn a remarkably effective cloak of secrecy. During the 1960s about 85 percent of the work of Fort Detrick's hundreds of researchers was classified, as were all of Dugway's tests.

Seymour Hersh has suggested that CBW "overclassification" was designed to avoid domestic disapproval rather than the enemy's gaze. CBW information was so closely held that even ranking Pentagon civilians and White House defense advisers understood little about it. Ironically, data were routinely disbursed to ten U.S. allies. "I always figured everything we sent them they passed on to Moscow," a recalcitrant former Chemical Corps officer told Hersh.

Decades-old documents are still kept secret. Research for this book included a Freedom of Information Act request for BW test data from the 1950s and 1960s. The army acknowledged the existence of at least 1.5 million pages of pertinent information, only a tiny fraction of which has so far been released.

HANDLING PROBLEMS

On March 13, 1968, the army was conducting a routine aerial nerve gas trial at Dugway when something went terribly wrong. The plane released persistent VX gas from too high an altitude. The cloud wafted more than thirty-five miles east, settling in nearby Rush and Skull valleys. Within days 6,000 sheep were dead. For months the army denied responsibility, but it finally admitted the accident in the face of intense press and congressional scrutiny.

This incident publicly and vividly revealed the army's callous recklessness with the health of its own researchers and that of the public at large. Moreover, it was not an isolated case. At least 221 people were exposed to chemical agents during munitions-testing accidents at Dugway from 1952 to 1969. Two fatalities were acknowledged. And in 1967 a helicopter pilot was overcome by a defective tear gas canister. The ensuing crash killed all 5 persons aboard.

Large tracts at Dugway remain a toxic graveyard, littered with unexploded CBW munitions, including a rusting 1,000-pound bomb whose contents are unknown. "The base will have to be off limits forever," a "high-level security officer" told a Salt Lake City newspaper in 1979. "A lot of that stuff is buried in the sand and will not deteriorate or neutralize for many years."

Despite such hazards, only part of Dugway's 175-mile border is fenced, and even fenced areas are easily penetrated. On many occasions "tourists, students from nearby universities, or other personnel" have entered test ranges by accident, a Dugway official acknowledged.

On three occasions in 1986 civilian contractors hired to do construction work at Dugway hit pockets of toxic chemicals that the army admitted had been buried at random years earlier. The men suffered severe headaches, nausea, fatigue, and liver dysfunction that lasted for months. Neither their doctors nor the army have identified with certainty the chemicals unearthed by the unlucky diggers.

Dugway officials also admit that on at least four occasions other than the sheep kill incident, toxic weapons, including nerve gas and the superhallucinogen BZ, were accidentally released outside the proving ground. And in one particularly bizarre case small bombs that were designed to be filled with BW agents but that instead were full of dye washed up on the shore of Carrington Island in the Great Salt Lake.

The passage of time has generated another critical public health question: How should these large quantities of obsolete CW agents and toxic substances be disposed of? In 1960 toxic chemicals generated at the Rocky Mountain nerve gas arsenal in Colorado were poured into a massive tunnel. A month later the region experienced its first earthquake in eighty years.

Initially this did not deter the disposal process. But five years, 165 million gallons, and 1,500 earth tremors later, the army reconsidered the wisdom of this practice. It concluded that it would be better if the noxious liquid were pumped back out. But a logistical problem arose: At the maximum feasible rate, the pumping would take more than 1,000 years. Presumably the deadly chemicals are slowly leaching into the water table.

During the past few years the army has been further refining disposal plans for old stocks of obsolete CW munitions. Incineration is the latest method of choice. But at every location considered as a burning site, the local residents, suspicious of the army track record, have objected strenuously. So far the destruction plans have remained stymied.

Biological research was conducted in a similarly cavalier fashion. The army claims only a handful of deaths resulted from its BW lab research at Fort Detrick, but that information is difficult to appraise in view of the fort's secrecy. Detrick had an agreement with the U.S. Public Health Service that no announcements would be made about lab workers stricken by BW-related diseases unless a worker died. Even then the cause of death was concealed.

According to a 1965 study by a Detrick safety officer, 3,330 lab accidents took place there from 1954 to 1962 alone, resulting in 410 infections and several deaths. Only a handful of such accidents have been forced into public view. In 1964 fifteen civilian employees of Detrick inhaled aerosolized staphylococcal enterotoxin B—milligram for milligram one of the most deadly agents ever studied. The accident came to light in 1982, when the wife of a permanently disabled victim obtained a classified report through Senator Charles M. Mathias, in an effort to obtain a damage award.

Operation Whitecoat subjects originally consisted of volunteer enlisted men. But officials switched to the more docile Seventh-Day Adventists, according to Hersh, after the soldier-guinea pigs had staged a sit-down strike to obtain more information about the dangers of the tests.

As indicated earlier, the BW *simulant* tests conducted in the 1950s and 1960s may have caused many illnesses. But by far the most dangerous feature of the army's BW program was live-pathogen testing.

"I was dressed in a rubber suit, high rubber boots, and rubber gloves," said a biologist who participated in BW field tests at Alaska's Fort Greeley. "I was instructed to touch nothing but the vegetation. . . . I noticed the carcasses of foxes, squirrels, rabbits, mice, weasels, owls, ravens, jays and small songbirds. . . . All that used to inhabit the enclosure was dead." The protective boundary he described was a seven-foot fence.

This kind of trial, whereby infected animals, birds, and insects could obviously escape from the site, was routine, particularly at Dugway. Outside the "test grid," fugitives were untrackable. The safety rationale was that only diseases that were endemic to the region were to be introduced. This practice, under any conditions, is a public health official's nightmare.

And careful review of available information shows that the army occasionally violated even this minimal standard by introducing foreign organisms. One of these was VEE. The army had studied VEE for decades, considering the highly infectious disease an ideal biological warfare candidate. A congressional hearing in 1969 revealed that animals on private farms near Dugway had been exposed to VEE, a disease never before seen in this country outside Florida and Louisiana.

Anthrax, which forms spores that remain potent indefinitely, was tested on a Dugway salt flat. The army had designated the test site a "permanent biologically contaminated area" but removed the designation after extensive testing had failed to reveal contamination.

Lieutenant Colonel Harold Hodge, a former commander of Dugway, disagreed with the change. "It is not possible to decontaminate anthrax spores," he said in 1979. "It lives in the soil and stays there—and is dangerous to humans as well as horses."

Animals that became infected during the field tests were checked regularly in the test area. The University of Utah, which monitored the spread of these diseases for Dugway, was also charged with creating contingency plans in the case of accidental "dissemination of a disease entity beyond the boundaries of a selected site," according to a university contract.

Apparently these contingency plans got off to a late start. In 1968—following nearly two decades of tests—a contract proposal to Dugway stated that monitoring had just begun near Salt Lake City. "We feel that it is in the best interests of the government to have baseline sampling data for areas near large centers of human population," the proposal noted.

Whether years of induced animal epidemics caused human illnesses or deaths may never be known. Secrecy, inadequate monitoring by the army, and the difficulties of tracking local public health records are by themselves formidable problems. Perhaps more important, the unpredictable migration of infected animals, birds, and insects makes it impossible to know how large an area to investigate. And once a disease is established in an animal population, it tends to form a permanent reservoir that can generate new epidemics.

THE ENDS AND MEANS QUESTION

Every government engages in subterfuge to protect what it perceives as legitimate national security interests. To this end, over several decades, the DOD and CIA have systematically lied about, hidden, and disguised the range, depth, and goals of their CBW enterprises as well as the national policies on which they were based. In the process they have corrupted public and private institutions, sacrificed unwitting research subjects, and ignored serious public health and safety concerns.

With few exceptions—such as the reluctant admissions following the Dugway sheep kill—CBW work of questionable morality, legitimacy, or propriety has been kept secret for years or decades. Whether "national security" has warranted such actions, particularly in peacetime, may be less important than using knowledge of this history to evaluate the military's current claims that its work is safe, open, and purely defensive.

3

AGENT ORANGE TO YELLOW RAIN: THE SPECTER OF MODERN CHEMICAL AND BIOLOGICAL WARFARE

BOLD STEPS FORWARD

"Biological weapons have massive, unpredictable, and potentially uncontrollable consequences. They may produce global epidemics and impair the health of future generations. . . . Mankind already carries in its hands too many of the seeds of its own destruction. By the examples we set today, we hope to contribute to an atmosphere of peace and understanding."

With these words, on November 25, 1969, President Richard Nixon unilaterally renounced the use and possession of biological weapons, ostensibly ending an effort of three decades. He ordered the destruction of America's BW stockpiles. He ended all but defensive research. Toxin weapons were added to the order three months later, and Nixon imposed a moratorium on the production of lethal chemical weapons.

Clearly these actions were significant in slowing the CBW race. Nixon's apparent ascent to the moral high ground was laudable statecraft. The move was not undertaken in a vacuum, however. He anticipated the signing of the 1972 Biological and Toxin Weapons Convention.

Despite serious limitations (assessed in detail in Chapter 7), the 1972 treaty represents the most advanced multilateral arms agreement ever signed. It is unique in banning possession and development of a class of weapons, rather than merely imposing limits on stockpiles or first use, as other accords dictate.

By 1969 most qualified military planners had concluded unequivocally that BW agents are a poor choice for theater operations. Backfires and accidents pose grave risks to an attacker's troops and friendly populations alike. The long incubation period makes BW unsuitable for gaining rapid tactical advantage. Because biological organisms might spontaneously mutate into more virulent strains, even immunized occupation troops would run some risk.

And BW is virtually useless for retaliation. "Few, if any, military situations can be imagined in which a state would try to redress a military imbalance by retaliating with weapons whose effects would not show up for days," said a U.S. disarmament negotiator in 1970.

BW and TW do have strong potential as weapons of sabotage and for clandestine attacks. These weapons elicit such profound revulsion, however, that using them could trigger severe political repercussions. In light of these factors, the practical military utility of BW and TW was seen as so narrow as to be inconsequential—leading to the signing of the BW convention.

Simultaneously the Nixon administration recognized strong incentives to show its commitment to arms control. By 1969 it was the only major power that had not yet ratified the 1925 Geneva CBW protocol, and it was taking a political and diplomatic beating over the Vietnam War. Nixon badly wanted to move in the direction of détente with the Soviets and to obtain an agreement on nuclear arms. BW disarmament represented a valuable confidence-building measure.

With no industrial lobby to push against the initiative for the sake of lucrative contracts, the protests of the army and its allies were overridden. As will be shown in later chapters, the advent of recombinant DNA has changed each condition that led to the 1969 renunciation and the 1972 treaty.

THE WORLD'S CHANGING CBW ARSENAL

After World War II most CW powers disarmed. By 1969 only three nations were believed to possess militarily significant supplies of chemical arms: the United States, the Soviet Union, and France. The French have never officially acknowledged the size of their stockpile—primarily nerve gas—but the United States estimates that it stands at about 400 to 500 agent-tons filled into about 5,000 tons of artillery munitions. This is sufficient to contaminate more than one-third of the surface of East Germany for up to several weeks.

Other Western nations accept the need for a modest CW deterrent arsenal, for which they depend on the United States. But they are uneasy about it. Europeans generally prefer to rely on protection-based tactics, such as impermeable garments and intensive decontamination training.

The approach has great strengths. Not the least of them is that it avoids the potentially destabilizing influence of a defense founded on massive chemical deployment at large, forward-based staging areas. The mere presence of these arsenals ensures similar preparations by opposing forces. Such conditions pose an ever-present risk that chemical battles could break out after a single impulsive action or accident.

A protection-based defensive strategy has just the opposite effect. It decreases the need for fully deployed chemical stocks by either side. And because a virtual tidal wave of chemicals would be required to overwhelm effective modern countermeasures, an aggressor would have to ready chemicals on an unprecedented scale in order to mount a successful attack. Such deployment would probably be detected far in advance, greatly reducing the element of surprise—essential for success against sophisticated defenders in the first place.

Only West Germany allows American CW on its soil, in approximately the amount of the French supply. The DOD has prepared a detailed plan for an emergency nerve gas airlift to Europe; that itself entails substantial logistical problems and public health dangers. One of these is the prospect that transport planes could be shot down by the Soviets.

Although little is known about Warsaw Treaty Organization CW capabilities, most analysts assume that WTO members relate to the Soviets in a manner similar to the NATO model. Western sources believe the Soviets have deployed CW in East Germany, Poland, and Czechoslovakia, although the Soviets deny this assumption.

The United States has steadfastly maintained that the Soviets have used the Nixon moratorium to enhance their CW stocks and to seize a decisive advantage. "While we stand on the high ground with our backs turned and our heads bowed, the Soviet butchers are gassing hundreds of thousands of people," said Representative Marvin Leath of Texas in a congressional debate over CW modernization. His comments, though more strident than most, reflect the level of acceptance U.S. claims have won among legislators and the public.

Recent American estimates about Soviet chemical stocks have ranged from 30,000 tons to an awesome 700,000 produced by eighteen to fifty production facilities. This uncertainty illustrates a fundamental intelligence problem—the absence of hard data. The Stockholm International Peace Research Institute points out that the most frequently quoted estimate—350,000 tons—is simply the average of the top and bottom of the range. Furthermore, SIPRI notes, there is no credible evidence that the Soviets have engaged in a CW buildup since 1969.

"We really don't know a thing—about how many weapons they have, what sort of delivery capability," commented one CIA analyst in 1984. "Anyone who says otherwise is kidding himself." The congressional General Accounting Office (GAO) reached a similar conclusion in 1983 after an exhaustive study.

Part of the argument may be semantic. The Soviets are said to have up to 85,000 troops dedicated to CW protection alone. But Saul Hormats, who once directed the U.S. CW program, says these troops "are essentially army janitors; they wash down the cannon or tank, but they do not fire it. . . . We have the identical kind of personnel and about the same number as the Soviets." The estimates may be further skewed by the multiple duties of Soviet CW defense troops, which also handle biological and nuclear decontamination.

Nonetheless, repetition of the charges has led to wide acceptance of the idea that the Soviet Union enjoys an overwhelming CW superiority. The Kremlin, admittedly, has done little to allay such fears. When it admitted in 1987 that it produces and stockpiles CW, it was the first Soviet comment on the subject since 1938. In any case, it would be surprising if the Soviets have ignored CW research and development when they know the Americans have not.

A DEVELOPING U.S. STANCE ON CW

Following Nixon's 1969 order, most U.S. CW production ended. This left a massive stockpile of mustard and nerve gases, other agents, and munitions stored in the continental United States, on Johnston Island in the Pacific, and in West Germany. Precise figures on the stocks are classified, but unclassified sources imply that the arsenal weighs in at slightly more than 40,000 agent-tons. Showing they are not above engaging in the kind of speculative hyperbole that characterizes U.S. estimates, the Soviets claim American holdings total 300,000 tons.

The Reagan administration has consistently decried the state of the CW deterrent, portraying America as virtually disarmed. By most other standards 40,000 tons of CW agents—as potent now as the day they were manufactured—are considered sufficient to give pause to potential aggressors.

According to a SIPRI analysis of U.S. sources, munitions containing 7,000 tons of CW agents are serviceable and ready to use. And 20,000 tons of CW agents stored in bulk could easily quadruple total usable munitions—a formidable amount, considering the potency of these weapons. VX, the most toxic nerve agent, for example, can kill at the 0.3-milligram level. This is equivalent to 3 billion lethal doses per ton.

Some of the chemicals are corrosive, however, and over the years have rendered a fraction of the munitions leaky or unreliable. Other munitions, such as the 105-millimeter howitzer shell, that were essential to field strategies of the 1960s are now obsolete or obsolescent. This is the crux of the debate that has raged for a decade. The administration has consistently pushed for renewed production and modernization of the chemical stockpile.

The 1969 moratorium on chemical production was accompanied by a diminished effort in research and development. This was dramatically reversed during the Reagan administration, as indicated by inflation-adjusted figures on U.S. spending in Table 1.

This work involves a wide range of attempts to develop more effective CW agents, with putative potencies 30 to 300 times that of nerve gas. Such agents, which have proved elusive so far, could relieve the problem of volume: the tons of bombs required to neutralize a large military target. Other efforts are focused on devising agents that effectively incapacitate without killing, which would burden enemy forces with debilitating medical logistics problems.

TABLE 1
UNCLASSIFIED BUDGET FOR U.S. CHEMICAL WEAPONS RESEARCH, DEVELOPMENT, TEST, AND EVALUATION (CONSTANT 1972 $ MILLIONS)

Fiscal year	1970	1973	1976	1979	1982	1984	1985
Amount	70	71	23	29	78	127	110

Sources: for 1970 and 1973: SIPRI. For 1976–1984: U.S. DOD (see source notes for details).

By far the most attention and money have been directed toward binary weapons, whose concept is based on loading two precursor chemicals into munitions. Only after the shell or rocket is fired do the chemicals mix to form nerve gas. The main rationale for binaries involves safety and flexibility. Ostensibly, two relatively innocuous chemicals, stored and transported separately, pose few of the extreme risks associated with deadly nerve gas.

Safer weapons, according to DOD logic, would not only protect public health but constitute a more convincing deterrent—easier to use and transport to hot spots. This all is based on the highly disputed assumption, of course, that the existing U.S. arsenal is inadequate.

From its first year in office the Reagan administration pushed hard for a binary program—to cost at least $2 billion—but was consistently rebuffed by a rebellious House of Representatives. Senate approval required Vice President George Bush to step into his constitutional role as president of the Senate in order to cast tie-breaking votes. The administration appeared to gain the upper hand, however, shortly after the Soviets had shot down Korean Air Lines Flight 007 in 1983; the House, fearful of being branded "soft on communism," authorized full-scale production of the weapons. Subsequent votes, taken after the airliner furor had subsided, stalled the program, but production of some forms of the weapons finally began in October 1986.

Congress also approved the construction of manufacturing facilities for (though delayed actual production of) the centerpiece of the program, the bigeye bomb, designed to shut down enemy staging and supply areas with VX. Not only is VX the most toxic nerve agent, but it has the viscosity of motor oil—sticking to everything. Because VX persists for days or weeks before dissipating, it can immobilize wide areas. Approval was given in spite of the fact that the bigeye has been plagued for two decades with a host of technical failures—from inadequate mixing of precursor chemicals to midair explosions.

The House allowed final assembly of other binary weapons to begin in October 1987, contingent on approval of force goals—long-term military plans for individual nations—by NATO's governing body, the North Atlantic Council. In May 1986, when NATO's Defense Planning Committee adopted these force goals, the Reagan administration considered the vote sufficient approval.

But NATO political leaders voiced firm opposition. Seven unequivocally rejected stationing binaries on their soil. Six others approved, in principle, an airlift from the United States during a crisis situation, but even these reserved veto power over any decision to deploy.

Given these objections, Representative Dante Fascell, chairman of the House Committee on Foreign Affairs, called the administration's acceptance of the NATO Defense Planning Committee approval "a farce from a legal standpoint and folly from a foreign policy and national security standpoint." Others accused the president of flouting a congressional mandate. The fate of binaries was thrown back into limbo. But in fall 1986 the Senate-House conference committee, working to resolve differences between the two versions of the binary funding bill, adopted the Reagan administration view accepting NATO Defense Planning Committee approval. Still, one more round of appropriations votes must precede final assembly of binaries.

ESCALATING DANGERS

Critics of binaries say they are far more destabilizing than other CW. The existing chemical arsenal is so dangerous that it is handled with the kind of caution usually reserved for nuclear weapons. This is part of the reason why chemical warfare has generally been considered militarily irrelevant since World War I. Chemicals have been seen as more trouble than they are worth.

"Safer" weapons are more easily integrated into force structure and strategy, generating concern that they could be used for first strikes. These fears are fed by the fact that a credible deterrent to Soviet aggression must be deployed close to the likely action—undoubtedly Europe—a deployment that nervous NATO allies may never permit. In any case, as we've seen, chemicals are of little value for retaliation. Attacking troops would be equipped with protective gear, but in adjacent areas civilians would suffer grotesquely. Such conditions could quickly provoke nuclear escalation.

"The production and potential use of chemical weapons in fact lowers [sic] the threshold of mass destruction," said Senator John Kerry during the 1986 defense authorization debate. "In the event of a chemical weapons exchange, there is no telling what the 'losing side' might do to avoid a defeat of such magnitude."

The United States is attempting to solve a perceived disadvantage with an offensive technological fix. But other means could be employed to bolster the existing deterrent. Uneven NATO defense capabilities could be upgraded, and conventional weapons could be used to achieve CW tactical goals. For example, runway cratering bombs, rather than VX, could effectively immobilize rear-area airports.

CW proliferation has also become a problem of major proportions since 1969. As discussed below, Iraqi attacks on Iran have profoundly affected world perceptions of these weapons. At least twenty-six other nations are thought to possess, or are seeking, a militarily significant chemical capability; they include Egypt, Syria, Israel, Vietnam, China, and Taiwan.

Proliferation stems partly from the ease with which CW can be produced from standard industrial chemicals. Nerve gas, for example, is based on organophosphorus compounds routinely used in the pesticide industry—believed to have been Iraq's path to obtaining nerve agents. And according to University of Sussex, England, CBW expert Julian Perry Robinson, the production of binary agents ironically requires less technical capability than the production of typical nerve gas.

In response, Western nations have imposed trade restrictions on precursor chemicals to countries suspected of seeking a CW capability. The ban is widely regarded as impotent, however. Any attempt to control world trade faces daunting challenges to begin with. And the credibility of CW export controls was badly damaged in 1986, when it was revealed that the United States, with the cooperation of Israel, had abrogated its own conventional arms embargo against Iran in order to obtain the release of U.S. hostages in the Middle East.

The importance placed on CW by the superpowers may be the primary engine for proliferation. Developing nations take very seriously the mounting perceptions of the Soviet threat and the intense U.S. push toward binary modernization. One effect, according to SIPRI, "might be to validate the otherwise questionable worth of CW armament and thereby to set a fashion which hitherto uninterested armed services around the world feel they ought to emulate."

MODERN BIOLOGICAL DEVELOPMENTS

By 1969 only the Americans and Soviets maintained significant BW offensive programs. In perfunctory statements after the 1972 BW treaty had been signed, the Soviets said they possessed no offensive capability. In contrast, the United States showcased destruction of its BW agents and the conversion of part of Fort Detrick to cancer research, part to "defensive" BW research. An aura of humane concern replaced the offensive drive toward "public health in reverse."

Then came the CIA toxin scandal. In 1975 hearings before a Senate committee the CIA admitted willful violation of Nixon's renunciation. The agency had maintained a major supply of toxins and the tools of spies needed to deploy them, such as precision dart guns.

The amounts held were astounding—enough to kill hundreds of thousands of people—particularly considering that the agency had never been authorized to conduct CBW actions in the first place. The agent that caused the greatest concern was paralytic shellfish toxin. The eleven grams on hand represented fully one-third of the total supply ever produced in the entire world.

"If the Director wishes to continue this special capability," wrote CIA Deputy Director Thomas H. Karamessines to the then CIA director Richard Helms after the presidential order to destroy the toxins, "it is recommended that the . . . existing agency stockpile be transferred to the Huntington Research Center, Becton-Dickinson Company, Baltimore, Maryland." All arrangements had been made, he noted.

In the heat of the highly publicized hearings the then CIA director William Colby was asked to verify that the agency was not hiding any other cache. "We obviously are conducting such investigations and releasing such orders as possible," he replied. "But I cannot be absolutely sure that some officer somewhere has not sequestered something." Not surprisingly the committee found the response less than reassuring.

Both superpowers maintain substantial "defensive" BW research efforts, permitted by the treaty. The most significant question about such research is that the difference between offense and defense is almost purely a matter of intent.

In 1969 National Security Adviser Henry Kissinger explained this point graphically in his Decision Memorandum 35, noting that Nixon's order "does not preclude research into those offensive aspects of bacteriological/biological agents necessary to determine what defensive measures are required." In other words, almost anything goes.

Correspondingly, U.S. research has covered everything from molecular genetics to munitions. Available information on Soviet efforts is sketchy, but the Soviets are surely not ignoring the questions considered important in the West. And the dangers do not stop at the superpowers' doorsteps. As with CW, the importance placed on BW research by the Soviets and Americans inevitably influences how smaller nations view the utility of such weapons.

THE DUBIOUS BODY COUNT

Modernization, stockpiling, proliferation, and, perhaps more important, increasing speculation about these trends have led to a dramatic rise in allegations of CBW use since 1969. Many of the examples are unlikely, unverifiable claims by antagonists in third world conflicts or part of the ongoing U.S.-Soviet propaganda war. Table 2 presents cases that have appeared in the international press or have been cited by arms control groups or government agencies.

TABLE 2
ALLEGATIONS OF CBW USE, 1969–1986

ATTACKER TARGET	YEAR(S)	AGENT(S)	NO. OF ATTACKS CASUALTIES	PRIME SOURCE(S)
A=Afghan rebels (Mujahedin)	1980–81	"Lethal chemical grenades"		Afghanistan/USSR
T=Afghan and USSR Troops	1984	CW	14 dead	USSR
	1986	CW/water poisoning	100 injured	Afghanistan
A=Angola T=UNITA rebels	1985–86	CW; napalm	3 dead, several injured	UNITA
A=Burma T=Shan rebels	1986	CW/2,4,D herbicide	Some deaths "extensive human toxicity"	Shan
A=Chad T=Chad rebels (TGNU)	1986	CW		TGNU

ATTACKER TARGET	YEAR(S)	AGENT(S)	NO. OF ATTACKS CASUALTIES	PRIME SOURCE(S)
A=China (PRC) T=Vietnam	1979	"Toxic gas," poisoning of water supplies		Vietnam USSR
A=Ethiopia T=Eritrean secessionists; Somalia	1980–82	CW	3,000 Eritrean casualties	Eritrean secessionists U.S.
	1986	Nerve gas		Missionary doctor
A=El Salvador T=FMLN/FDR rebels and civilians	1980–84	"toxic gas," acid spray		FMLN/FDR USSR
	1984–85	BW; white phosphorus	Many	Mexican press Salvadoran doctor
	1985	Sulfuric acid		A refugee
A=Great Britain T=Argentina	1982	BW		Argentine press
A=Guatemala T=Rebels	1982	BW		Cuba
A=Iran T=Iraq	1984	CW; tear gas		Iraq
A=Indonesia T=East Timor rebels FRETILIN	1984	CW		FRETILIN
A=Iraq T=Iran	1980–86	Mustard, nerve gas, vesicant	Attacks: more than 175, Dead and injured more than 3,500†	Iran*
	1984	BW/anthrax TW		Israel
A=Israel T=Lebanon	1982	CW		WAFA
A=Laos T=Laotian rebels (Hmong)	1975–83	TW; mustard; nerve gas; irritants	More than 260 attacks, More than 6,400 dead many injured	U.S.

ATTACKER TARGET	YEAR(S)	AGENT(S)	NO. OF ATTACKS CASUALTIES	PRIME SOURCE(S)
A=Nicaragua T=Nicaragua rebels (NUDF)	1984 1985	TW CW		NUDF NUDF
T=Nicaragua rebels (KISAN)	1986	CW		KISAN
A=Philippines T=MNLF rebels	1984	CW		European press
A=Portugal T=Colonies Angola Mozambique Guinea-Bissau	1969–74	CW; BW; chemical herbicides		U.S. European press (herbicides)*
A=South Africa T=Angola	1978, 1982	"Paralyzing gas"		United Nations*
A=South Africa UNITA rebels T=Angola	1984	CW		Angola
A=South Africa T=SWAPO rebels	1984	CW/herbicides		USSR
A=Sudan rebels (SPLA) T=Sudan	1986	CW		Sudan
A=Thailand T=Kampuchea	1984–85	CW		Kampuchea
A=USSR T=Afghan rebels (mujaheddin)	1979–82	TW; nerve gas; irritants	More than 83 attacks, More than 3,342 dead, many injured	U.S. Mujaheddin
	1985	CW		U.S.
A=US T=Vietnam	1969–71	CW/herbicides; irritants	Millions of crop acres; many birth defects, cancers, etc.	Many*

ATTACKER TARGET	YEAR(S)	AGENT(S)	NO. OF ATTACKS CASUALTIES	PRIME SOURCE(S)
	1969–70	BZ		Dutch author
A=U.S. T=Cambodia	1970	Nerve gas		Western press
A=U.S. T=Cuba	1971	BW/swine fever	More than 500,000 swine,	Cuba, USSR
	1978–81	BW/sugarcane rust, tobacco mold, swine fever, hemorrhagic dengue and conjunctivitis	Major crop loss More than 173,000 swine; 156 dead; More than 344,000 ill	Cuba, USSR CIA, U.S. press
A=U.S. T=Pakistan, India	1972–84	BW (tests)/yellow fever, cholera	More than 30 dead	USSR Cuba
A=U.S. T=Grenada	1983	CW		USSR
A=U.S. T=Brazilian Indians	1984	CW/herbicides**	More than 7,000 dead, Many birth defects	USSR
A=U.S. T=Honduras	1985–86	CW	Many injuries	Honduran trade union
A=U.S. T=Nicaragua	1985–86	BW/dengue; hemorrhagic dengue	More than 500,000 cases	Nicaragua
A=Vietnam T=Kampuchea rebels (Khmer Rouge and KNPLF, NADK)	1978–85	TW; CW/irritants, cyanide nerve gas; phosgene	150 attacks, 1,046 dead	U.S., Khmer Rouge, Thai Army
	1984	CW incapacitant		KPNLF
	1986	BW	179 casualties	KPNLF
	1986	CW	219 casualties	NADK
A=Vietnam T=China (PRC)	1979	"Poison gas"		China, U.S.

ATTACKER TARGET	YEAR(S)	AGENT(S)	NO. OF ATTACKS CASUALTIES	PRIME SOURCE(S)
A=Vietnam T=Thailand	1982	TW	4 attacks	U.S.
A=Zaire/ Shaba rebels T=Zaire/ Shaba rebels	1977	Poison arrows		Zaire, Shaba Rebels

* Independent confirmation beyond reasonable doubt. All other reports are strictly unproved allegations, many of which have been denied by accused perpetrators.
† SIPRI estimate. Iran estimates 12,600 casualties.
** This involved a deforestation program defined by USSR as CW testing.
CW. Specific chemical agent unspecified or unknown.
BW. Specific biological agent unspecified or unknown.
TW. Trichothecene mycotoxins.
Country name refers to government troops or sources.

Primary sources: U.S. government, SIPRI, worldwide press reports (see source notes for details).

A few of these cases were openly acknowledged by the attacker or verified by independent observers. Most are hotly disputed. Regardless of their authenticity, these claims vividly demonstrate the methods and rationale for using CBW today and in the future.

THE CASE OF THE VIETNAM WAR

During the Vietnam conflict the United States unleashed against the North Vietnamese and Vietcong guerrillas a volume of chemicals not seen since World War I. From 1965 to 1971 more than 10,000 tons of "harassing" or "riot control" agents—primarily CS tear gas—were used for counterinsurgency. In addition to severe and painful irritation of the eyes and respiratory tract and temporary blindness, CS can induce blistering and violent vomiting.

"We think that the use of tear gas is more humane than other weapons," Secretary of State William P. Rogers told the Senate Committee on Foreign Relations in 1971. "It protects in many instances the lives of the enemy [and] American lives." Rogers's point was that by using nonlethal CS to flush enemy troops out of hiding places, they could be captured rather than blown up, while GIs would face fewer snipers and ambushes. In fact, gas was rarely used for such high-minded goals.

"Occasionally, the use of riot gas . . . has led to the capture of enemy soldiers who otherwise might have escaped or been killed," Harvard's Matthew Meselson told the committee. "But these very limited occurrences are greatly overbalanced by the use of CS to enhance the lethal effectiveness of conventional fire power." Translation: Gas them out; then shoot them. In any case, "nonlethal" CW is a misnomer. In high concentrations any war gas or weaponized microorganism can be deadly.

During Operation Ranch Hand (1961–71) the U.S. Air Force dropped tens of millions of gallons of chemical herbicides on Vietnam. "If you are traveling down a particular highway you can use herbicides to destroy the foliage alongside to prevent ambushes," Rogers explained in 1971. He neglected to mention that the air force laid to waste up to an estimated 44 percent of the nation's upland forests. Neither did he include crop destruction—a U.S. method of warfare during every armed conflict in our history—which was very effectively used on 8 percent of all Vietnamese agricultural land.

The herbicides chosen for Operation Ranch Hand were code-named Agents Blue, White, and Orange. The infamous Agent Orange is a mixture of 2,4,D and 2,4,5,T—chemicals banned in the United States and other nations. The former is a potent carcinogen and teratogen (causing birth defects), while the latter is inevitably tainted with dioxin, one of the most toxic substances that exist. According to the Environmental Protection Agency (EPA), concentrations of dioxin in water higher than 2.1 parts per quintillion (1 followed by 18 zeros) constitute an unacceptable danger to human health. It was dioxin contamination that led to the tragic dissolution of the town of Times Beach, Missouri, in 1983.

The spraying inflicted incalculable ecological and economic devastation and human suffering on the Vietnamese. Its American legacy was also horrendous: Tens of thousands of Vietnam veterans and their offspring suffered cancers and a range of other serious health problems, apparently linked to Agent Orange exposure.

During the Vietnam War the United States was the only major power that had not yet ratified the 1925 Geneva Protocol on CBW, although it claimed to adhere to the pact. By U.S. definition, herbicides and harassing gases fell outside treaty provisions—a premise accepted by only Australia and Portugal at the time.

This position, which has great bearing on current U.S. CBW programs, is addressed in later chapters. Spurred by outrage from enemies and allies alike, the United States phased out the chemical war in 1971. In 1976 President Gerald Ford renounced first use of herbicides and restricted employment of riot control agents. These limits, however, maintain the "humane use" fiction and are so loosely worded as to limit only obvious, large-scale chemical offensives.

THE CASE OF IRAQ

When Iraq invaded Iran in 1980, it had the look of a winner. Saddam Hussein's modern army quickly seized coveted border areas from Iran's apparently confused, preoccupied revolutionary regime. Victor soon became vanquished. Iran went on the offensive, routing Iraqi troops to regain its territory. It then established footholds inside Iraq, including the strategically vital Fao Peninsula near Kuwait. Iran owed this dramatic success to a tactic not seen in many years: massive human waves.

Iranian men, and even boys, volunteered for battle by the hundreds of thousands, intoxicated by religious zeal and a thirst for martyrdom, willing to suffer frightful consequences. The Iraqis obliged. Increasingly desperate with each new Iranian thrust, they unleashed a series of CW attacks. Of more than 175 alleged attacks, at least 30 have been confirmed independently. Iran claims it incurred more than 12,600 gas casualties. Attacks with both nerve and mustard agents were affirmed by a United Nations inspection team in 1984. It is the only authoritatively verified use of nerve gas in history and the first involving mustard gas since World War II.

The course of the war has been influenced by many factors, but limited CW attacks and the threat of greater ones dampened the Iranian "final offensive" in 1985. More significantly the Iraqis provided a lesson the world was loath to learn, notes Brad Roberts, an analyst at the Georgetown Center for Strategic and International Studies: "Military planners may come to believe that in certain conflict scenarios, such as those in swampy terrain against an ill-equipped but highly motivated opponent, chemical weapons have some utility."

For Iraq, the desperate fight for survival outweighed the risk of political ostracism. How serious was that risk? Apparently relatively trivial. Because of near-universal suspicion of the militantly Islamic Iranians, few countries have even suggested political or economic steps that might reduce Iraq's fighting ability. The United States renewed diplomatic relations with Baghdad in 1984, ironically coinciding with the UN denunciation.

"Politically, the Iraqi use of CW poses great challenges to the non-aligned movement. If the political costs of using CW are seen as minimal, and as affordable, the military incentives for chemical weapons would multiply globally," concluded an Indian defense analyst. "Once the CW spread and are seen as legitimate, the advanced and interventionary powers . . . would most certainly use them in their conflicts with the Third World."

THE CASE OF CUBA: COVERT WAR

"With at least the tacit backing of U.S. CIA officials . . . anti-Castro terrorists introduced African swine fever virus into Cuba in 1971," noted a 1977 *Newsday* (New York) article based on interviews with intelligence sources. "Six weeks later, an outbreak of the disease forced the slaughter of 500,000 pigs to prevent a nationwide animal epidemic." The UN Food and Agriculture Organization labeled the outbreak the "most alarming event" of 1971. The highly contagious, usually deadly swine disease halted Cuban pork production for several months.

The *Newsday* report validated claims made by Cuba for years: that the outbreak was linked to ongoing, covert BW destabilization by the United States against its tiny neighbor. The campaign allegedly targeted Cuba's economic survival crops—tobacco (blue mold) and sugar (cane smut) as well as people—most notably an outbreak of hemorrhagic dengue, which affected 350,000, killing 156.

Cuba's charges could not be proved, but they were not farfetched. As noted above, the United States has frequently destroyed enemy crops. The CIA tried to overthrow the Cuban government at the Bay of Pigs and mounted numerous assassination attempts against Castro. Agency *interest* in sabotaging the Cuban economy with chemical or biological methods against crops and workers is also a matter of public record, as noted in Chapter 2. And Cuban investigators cite another piece of circumstantial evidence in the African swine fever case: The disease broke out simultaneously in two distant locations.

THE CASE OF YELLOW RAIN

In 1979 a tribesman from the Hmong hill region of Laos told Southeast Asia researcher Stanley Karnow a chilling tale: Two aircraft had sprayed a mysterious substance over his village. "I had a headache and my eyes swelled up, as if there were sand in them." Dozens of others suffered long bouts of diarrhea and vomiting. Days later they died, he said. This account, published in the *Washington Post,* was the initial mention in the U.S. press of the alleged toxin weapon "yellow rain."

More grisly stories soon followed. Victims were said to bleed from the eyes, nose, and mouth, to vomit and defecate blood. When they died, flesh peeled from their bones. The U.S. State Department claimed that over a period of years the Soviets in Afghanistan, and (at Soviet direction) the Vietnamese in Kampuchea, and the Laotians against rebellious Hmong tribespeople, killed more than 10,000 guerrillas and their supporters with yellow rain, allegedly a weapon derived from trichothecene mycotoxins, poisons produced by a fungus from the genus *Fusarium.*

Yellow rain became an American obsession. To the United States it was an open-and-shut case: The Soviets and their clients perpetrated monstrous attacks that flagrantly violated international law. But to critics of this assumption—enemy and allied governments, independent U.S. scientists and arms control analysts—the U.S. conclusion was fatally flawed. The strident American accusations, they said, were a monumental error in international relations.

The substantial arms control and treaty implications of what was to become one of the most complex and confusing diplomatic problems in decades are addressed in later chapters. The tale of yellow rain—indeed, the detective story—can fill whole books. Following is only a brief sketch of the main issues.

The State Department case was based on several factors:

- Physical evidence. Two Soviet gas masks contaminated with mycotoxins were obtained in Afghanistan. In addition, six samples of yellow rain on rocks, leaves, and twigs gathered at sites of apparent attacks were found to contain mycotoxins, as did blood or urine specimens from 20 of 100 alleged victims. The specific combination of trichothecene mycotoxins found in some samples could not occur naturally, U.S. experts claimed.

- Victim testimony. Hundreds of interviews with alleged witnesses and victims yielded voluminous information about attacks, illnesses, and deaths as well as symptoms that fitted the general description of mycotoxin poisoning.
- Defector testimony. The State Department cited Soviet defectors who claimed to have personal knowledge of TW activity in Afghanistan and received similar reports from Laotian defectors and eyewitnesses.
- Intelligence data. The State Department claimed that this *classified* information proved its case. These data were obtained through "national technical means"—a term that refers to satellites, reconnaissance aircraft, and a vast array of ground intercept stations and intelligence ships.

A host of independent scientists, led by Meselson, noticed gaps and inconsistencies in the U.S. case. Conducting their own investigation, they concluded that yellow rain is not a weapon but the product of a very natural phenomenon. The following is a synopsis of their view:

- Defector testimony and intelligence data. Neither the word of defectors whose identities were protected nor secret intelligence information could be corroborated and is therefore of little significance. "It would be wrong to pay attention to statements of this kind," a SIPRI report commented on the issue. "Matters of international law must be judged on the basis of evidence presented."
- Physical evidence. Despite seven years of intensive searching, not a single dud munition, spent shell casing, spray apparatus, or other sign of dissemination was found. The gas masks were obtained indirectly from questionable sources.

  As for attack site samples of yellow rain, of about 100 samples, only 6 showed the presence of mycotoxins—in trace amounts. Of these 6 positive findings, 5 were tested in a single lab and were not or could not be confirmed, casting doubts on their validity. Neither Canada nor Great Britain, which also analyzed battlefield samples, could find any evidence of mycotoxins. Environmental toxin experts discovered that the combination of mycotoxins found in yellow rain does occur in nature, contrary to U.S. claims.

  In 1982 investigators at the British Chemical Defense Establishment discovered that all the samples showed the presence of the same distinctive component, but it was not mycotoxins. The substance was pollen.

"Like yellow rain," Meselson and his colleagues noted in a 1985 article, "[honeybee] feces take the form of small, yellow, pollen-filled spots that dry to a powder." Many species of bees defecate in swarms from altitudes far beyond the range of the naked eye. After months of field research into bee "cleansing showers," both in the United States and Southeast Asia, followed by extensive analysis of Afghanistan battlefield samples and bee feces, they concluded that yellow rain and bee feces were one and the same.

Perhaps more important, at the concentrations found in environmental samples, said Saul Hormats, former director of the army chemical weapons program, "3,000 tons of light, fluffy toxin-containing material would have to be sprayed over a one-square block target" to yield the horrific consequences recounted in State Department publications.

The blood and urine testing, according to an analysis in *Chemical and Engineering News* that won a major science-writing award, was conducted with such lax methodology that the positive results would be considered unacceptable in any academic laboratory. Mycotoxins routinely infect grains, the report added. Therefore, even if toxins were present in the samples, their origin may have been contaminated food.

- Victim testimony. The bulk of testimony cited by the State Department came from Hmong tribespeople. Their suffering seemed genuine, but their recollections were inconsistent. Hmong statements about the color and appearance of yellow rain, and the method of dissemination, varied strikingly—even in connection with a single attack. Hmong descriptions of symptoms were similarly diverse, and only 25 percent of villagers at any alleged attack site became ill.

The Hmong, who worked closely with the CIA during the Vietnam War, are largely destitute refugees now, as the result of their unsuccessful attempt to overthrow the socialist Laotian government in the 1970s. When U.S. government investigators approached them, the Hmong could logically have perceived compelling immigration-related incentives for supplying yellow rain stories. In addition, the interviewers violated fundamental rules of survey research by making their interests known in advance, asking leading questions, and failing to select refugees randomly from the same villages in order to confirm stories.

And in 1987 Meselson and two colleagues obtained U.S. documents that acknowledged that some U.S. interviewers had failed to separate eyewitness accounts from hearsay.

"More subtle factors also limit the investigator," according to Jacqui Chagnon and Roger Rumpf, experts on Laotian culture who are fluent in Lao. "Story telling is an art . . . [and] truth stretches easily into exaggeration. . . . The Lao language possesses a vocabulary which is imprecise and technically limited. Discussions about 'poisons,' for example, can easily bring on a linguistic migraine." Their interviews in Laos revealed a profound confusion among the Hmong between yellow rain and Agent Orange, which the U.S. military had sprayed on their territory as recently as 1974.

The overall weight of the argument disputing State Department claims has persuaded virtually all other nations and independent scientists that yellow rain is not a toxin weapon. In an interview in November 1986 Meselson said that high-level sources in the government told him that during the preceding month top U.S. officials had finally been informed by their own scientists that yellow rain never existed.

The real residue of yellow rain is its impact on world perceptions of CBW. The U.S. position may be the result of intelligence failure, intellectual dishonesty, or calculated disinformation. Regardless, it has been thoroughly discredited. The State Department has shown no sign that it will retract or soften the yellow rain allegations or present the secret intelligence data that it says constitutes incontrovertible evidence. This Catch-22 position inevitably casts doubt on other U.S. claims related to chemical and biological warfare.

MODERN CBW PATTERNS

These recent CBW attacks share a striking similarity. All allegedly took place in the third world, and for good reason: The attractiveness of these weapons is directly related to the technological backwardness of the target. Developed nations have escaped for a complex range of reasons. One important factor is firepower; a massive conventional or even nuclear strike could be used to respond to a CBW attack. And other measures of preparedness are even more important in the success or failure of a CBW first strike.

Unlike the developed world, most developing nations lack sophisticated public health apparatuses. A CBW attack would cause far greater damage in countries with an absence of adequate supplies of vaccines and drugs and advanced medical delivery systems. The dissemination of

CBW and its overall impact would also be enhanced by primitive sanitation and water purification systems.

The Nicaraguan government blamed the CIA for a devastating epidemic of dengue fever during 1985 and 1986 that struck hundreds of thousands of Managuans. Whether the source was the CIA or nature, the epidemic raged for many months before subsiding—a result in large part of the primitive conditions of that nation. And massive herbicide intoxication suffered by the Vietnamese in the 1960s and 1970s was due, in part, to contamination of the water supply.

By the same token, poor countries are much more subject to economic destabilization. Cuba and Vietnam, for example, can be considered prime targets for crop destruction and animal epidemics, as suggested in the above examples. Monoculture—the planting of a single crop over wide areas, which is the norm in the third world today—is particularly vulnerable to CBW. Often a small nation's economic survival is wagered on one product (sugar in Cuba, coffee in Nicaragua). A combination of BW-induced epidemics and CBW agricultural sabotage could overwhelm limited health and welfare services, cause living standards to plummet, drastically affect morale, and induce financial desperation in a small nation. Ideally, from the attacker's perspective, this would foment political crisis, as suggested in this book's opening scenario.

SIPRI believes that a small island nation such as Cuba might be particularly vulnerable to a BW attack designed to soften its defenses for a conventional invasion. An agent that incapacitates but is only rarely fatal, such as VEE, might be a good candidate. Aircraft spraying or food and water supply contamination of militarily significant targets would precede military intervention by a few days.

After the organism's incubation period had elapsed, stricken troops would pose only weak resistance. Casualty figures would probably be lower than would have been likely in a conventional invasion. Pro-U.S. elements in the population would therefore not be alienated by extraordinary loss of life.

As the Italian attack on Abyssinia in the 1930s and the contemporary Iran-Iraq War show graphically, ill-equipped and poorly trained forces can be dealt devastating blows on the battlefield with CW. In contrast, modern armies with adequate protective gear would survive and respond to all but the most sustained chemical attack. This logic is not lost on nations that have employed chemicals in the past. For example, Egypt reportedly used CW in Yemen while intervening in the

1963–67 civil war; despite abundant preparations, however, Egypt refrained from using chemicals against the technologically advanced Israelis in 1973.

The need to maintain secrecy, or at least deniability, also favors use of CBW in the developing nations. For all forms of CBW, a lack of detection and evaluation equipment may render a victim unable to prove that an attack has actually occurred—an essential component in mobilizing international support and sympathy.

This applies most clearly to biological organisms. Diseases that are either endemic or could plausibly have been transferred to the target area can be chosen. For example, some analysts have blamed the 1971 epidemic of African swine fever in Cuba—the first recorded outbreak in the Western Hemisphere—on tainted food brought in by commercial airliners or merchant seamen.

Only the elusive smoking gun can actually prove BW allegations, and even preliminary evidence of BW demands sophisticated epidemiological studies and intensive laboratory work. Most developing nations lack such capabilities.

CW attacks are far easier to prove, for of course, it would be impossible to claim nerve gas could be of natural origin. And chemical injuries, such as the seared lungs and giant blisters associated with mustard exposures, are obvious to the trained eye. As the Iranians have shown, casualties can be transported out of battle zones for examination by independent authorities, leaving little room for plausible denial from the attacker.

But short of such obvious evidence, complex scientific analysis of environmental samples—often subject to skepticism about sources and methodology—is required. And in remote or particularly primitive regions, such as areas of Afghanistan, witnesses of CW attacks may be able to contact the international press but are often ill equipped, politically or logistically, to obtain credible proof. Deniability is thus preserved.

Toxins occupy the middle ground. As revealed significantly in the yellow rain controversy, toxins, like BW agents, exist in many natural environments. Collecting evidence in remote war zones is in itself difficult. Proof of a toxin attack is also highly dependent on expensive technoloy. Toxin warfare, again, is far more likely in the third world, where the scientific expertise and equipment necessary to test environmental samples are inaccessible, than in the developed world. The chief difficulties the State Department encountered trying to prove its yellow rain case involved sampling and analysis. When samples were obtained, only a handful of investigators, even in the United States, had the equipment and know-how to test them.

In view of these conditions, which are conducive to CBW methods, the potential range of future applications can be estimated:

- CW. Chemicals will find their role primarily as a weapon of desperation by faltering third world regimes against domestic insurgents or foreign invaders. They may also be chosen for counterinsurgency in remote, militarily sealed-off areas where the target group could not readily communicate its plight to the outside world.

  CW has often been referred to as the "poor-man's atom bomb." Effectively disseminated, nerve agents can neutralize an area three times as large as a comparable payload of high explosives. They are also cheaper to produce than nuclear weapons and do not require so formidable a scientific and industrial base. For example, Egypt has been accused of preparing a strategic CW capability to counter a supposed Israeli nuclear threat.

  CW may also be an attractive "halfway house" for nations looking for nuclear capability, according to Roberts. They provide an opportunity to develop and deploy delivery systems, a command, control, communications, and intelligence apparatus, and defensive warning systems—all essential to nuclear potential.

  At this point, however, the technology to deploy effectively the hundreds or thousands of tons of CW agents needed for a genuine strategic threat appears beyond the reach of all but the major powers. And a strategic CW threat does not fit into the realistic military applications of most third world nations: counterinsurgency and fighting off attackers on home soil.
- TW and BW. TW, sharing physical qualities with CW and deniability with BW, would most likely be employed for counterinsurgency. Third world nations presently lack the capability to produce and deploy militarily significant amounts of toxins, although they may be helped by major powers, as suggested by the United States in the yellow rain affair.

  Because their potential for mass destruction is exceeded only by nuclear weapons, BW agents have also been suggested as a nuclear replacement. But the reach of small or poor nations is local, at most regional. Large-scale use of BW would pose unacceptable risks for any attacker at close quarters. These risks would be far larger for a technologically backward nation.

Biological organisms, with their high secrecy-deniability factor, would most likely be chosen by major powers to disrupt unfriendly governments in the third world or put down insurgencies there. Economic destabilization would be the primary goal. New biotechnologies, as will be shown in later chapters, may be pushing this goal to the top of the military agenda.

4

POROUS BARRIER: CHEMICAL AND BIOLOGICAL WEAPONS TREATIES

THE LAW COMES OF AGE

For more than a century nations have attempted to legislate limits on the horror of unconventional weapons. The Brussels Conference of 1874 was the first in a series. This was a primitive era for the biological and chemical sciences. The ways chemicals and biological organisms act in the environment and affect the human body, and how these agents might be weaponized, were still largely unknown.

The language of the Brussels Declaration, though scientifically unsophisticated, expresses clearly the general dictum: "The laws of war do not recognize in belligerents an unlimited power in the adoption of means of injuring the enemy. . . . According to this principle are especially *forbidden* employment of poison or poisoned weapons." (Emphasis in original.)

The First International Peace Conference, held at The Hague in 1899, accepted the Brussels intent and sharpened its language, adding: "The contracting powers agree to abstain from the use of projectiles the object of which is the diffusion of asphyxiating or deleterious gases." All the great powers of the day except one ratified the treaty—a remarkable document that anticipated actual development of gas bombs.

The United States alone rejected it. That decision was justified by the technicality that weapons could not be banned until they were developed and by the possibility that gas could be more humane than bullets and bombs. The latter rationalization was to become a twentieth-century American refrain.

Although these limits were ratified by all the European countries involved in World War I, they did not stop the massive use of poison gas. The debacle of gas warfare taught the negotiators a tragic lesson about legal wording and good faith: When the Germans released the huge clouds of chlorine that wafted over their unsuspecting enemies at Ypres, technically they did not break any international law. The gas had been unleashed from cylinders and carried by the wind. No *projectile*—forbidden by The Hague treaty—was used. Later, formal violations were the product of the gas war's escalation.

Armed with refined legal vigilance, the victors compelled the defeated Germans to sign the following provision of the 1919 Versailles Treaty: "The use of asphyxiating, poisonous, or other gases and of analogous liquids, materials or devices being prohibited, their manufacture and importation are strictly forbidden in Germany." Similar treaties committed Germany's wartime allies Austria, Bulgaria, and Hungary to CW disarmament.

Images of wartime gas casualties stimulated substantial public opposition to CW and facilitated broader legal initiatives. The 1922 Washington Treaty on Submarines and Noxious Gases contained strong prohibitions against *using* CW, modeled on the Versailles language, although it contained no restrictions on *possession*.

Political maneuvering by President Warren G. Harding, his secretary of state, and Senator Elihu Root, who represented the United States in negotiations, helped ease the treaty through the Senate without a single dissenting vote. It never came into force, however, because the pact required ratification by the French, who objected to the provisions on submarines.

Still, the United States pressed forward for a CW agreement, and on June 17, 1925, thirty-eight nations signed the Geneva Protocol—a step of profound importance for world peace. The agreement became an emblem of the mutual aspiration that has promoted widespread restraint on chemical and biological warfare to the present day.

Not only did the protocol include the Washington Treaty ban on the use of CW in any form, but at the urging of the Polish delegation, biological weapons were added. (Hence the somewhat cumbersome title, "Protocol for the Prohibition of the Use in War of Asphyxiating, Poisonous or other Gases, and of Bacteriological Methods of Warfare.") The protocol stood for fifty years as the primary international biological arms control agreement and is still the major legal sanction against chemical attack.

Given the unanimous support for the Washington Treaty only three years earlier, President Calvin Coolidge, Harding's successor, grew complacent. He was caught totally unprepared for the onslaught of lobbyists that descended on the Senate when it took up the Geneva Protocol. The Army Chemical Warfare Service mobilized a battalion of economic and military interests, including the American Legion, the Veterans of Foreign Wars, and the American Chemical Society (ACS) and other industry groups.

The protocol, said the ACS leader, "meant the abandonment of humane methods for the old horrors of battle." Ratification of the Washington Treaty, the lobbyists argued, was produced by postwar chemical "hysteria."

The Chemical Warfare Service, with its esprit de corps and instinct for survival, won a major battle. Sensing defeat on the horizon, the Senate majority leader pulled the treaty from consideration. A full fifty years later, in 1975, the United States finally ratified, becoming the last major power in the world to do so. (By 1986 108 countries were parties. See Appendix 3 for the treaty's text and implementation status.)

In more recent times Pentagon lobbying has continued to wield powerful influence on CBW policy, despite explicit sanctions against use of military funds "in any way, directly or indirectly, to influence Congressional action on any legislation or appropriation matter before Congress."

In 1986 the congressional General Accounting Office and the chairmen of the House committees on Government Operations and Foreign Affairs accused the Defense Department of violating this statute by hiring lobbyists and public relations consultants in an effort to urge renewed CW production.

GENEVA PROTOCOL: STRENGTH WITHIN FRAILTY

Any arms control treaty has three modes of operation: legal, political, and moral. Unfortunately the Geneva Protocol's legal strictures are the weak leg of the stool. The pact does not in any way limit research, development, testing, or production. Horribly destructive weapons, the use of which is expressly forbidden, can still be stockpiled freely in any quantity. Thus the agreement indirectly encourages the buildup of massive deterrent arsenals.

Many parties to the treaty have reserved the right to retaliate in kind and within the obligations of their strategic alliances. This means that if the Soviets attacked the British with mustard gas, Britain could respond in kind. The United States, Britain's NATO ally, could also strike the Soviet Union with mustard gas. Nor would it be illegal as well for the Americans or British to attack East Germany, the Soviets' Warsaw Treaty Organization (WTO) ally, with mustard gas. Presumably such a potentially destabilizing response would be taken only in extreme circumstances, but technically it would not be a treaty violation.

The treaty is also silent about retaliation that is itself an escalation in the level of conflict. For example, if attacked with mustard gas, would Britain be able to respond, legally, with nerve gas? Military imperatives would most likely fill this legal vacuum.

The protocol is further hampered by limited acceptance. Some former colonies have never acceded to ratifications made in their behalf by former rulers. Other states, such as Afghanistan, Burma, and Honduras, simply have not signed the pact. Nonparty nations are legally capable of, and subject to, CBW attack by other nonparties. And even many strong supporters of the protocol reserve the right to use CBW against nonparties.

The pact also ignores technology transfer. CBW powers may pass these weapons to nonparties without violating the protocol. Allegations that the Vietnamese used Soviet-supplied toxins in the 1970s suggests exploitation of this loophole. Vietnam did not ratify the agreement until 1980.

Furthermore, the treaty leaves the question of verification completely open. It established no machinery for consultation or cooperation between parties in the event of an alleged violation. Without enforcement procedures the treaty regime is powerless to impose penalites or sanctions of any kind if an abrogation is proved beyond reasonable doubt. The unpunished Iraqi CW attacks on Iranian troops are a case in point.

These problems are not merely the blundering oversight of those who drafted the protocol. "Airtight" language is unrealistic and probably impossible. Rather, the authors sought broad, simple language that could be agreed upon even by political antagonists, rather than try to second-guess every conceivable contingency. Nations that wish to violate an arms control treaty will do so. Ypres is the classic example of the extent to which any international agreement relies to a great degree on good faith.

Not surprisingly, this broad language has led to varying interpretations. For example, the parties seem to agree that incendiaries, including the notorious napalm used by the Americans against the Vietnamese, are permitted. This is not the case, however, with chemical herbicides and harassing gases, such as CS, which were nonetheless used by the United States in Vietnam and are still employed by police forces around the world to quell domestic unrest.

As noted earlier, the United States views CS as a relatively "humane" weapon. This view has been widely disputed by other nations and by American experts in international law and was clearly contradicted by the Vietnamese experience. But as a sovereign state the United States is, of course, free to adhere to definitions of its own choosing.

In 1969 the UN General Assembly, in a symbolic action, voted 80 to 3 in favor of a resolution that includes *all forms* of chemicals within the protocol's purview. The three dissenting votes came from the United States, Portugal, and Australia, all of which had used or allowed the use of tear gas or herbicides during the 1960s. Thirty-six nations, mostly U.S. allies, abstained. Some of these have since adopted the U.S. view on CS.

In 1976 President Ford renounced first use of herbicides and placed some limits on the most aggressive applications of harassing gases. But even this did not affect the U.S. perspective that the Geneva Protocol does not apply to these chemical agents.

"This situation is dangerous," according to a SIPRI analysis, "not because tear gases or herbicides in themselves present any abnormal threat to international security, but because no unambiguous distinction can be drawn between these agents and the other 'poisonous . . . materials' specified in the Protocol: their legitimation therefore risks impugning the entire body of law that stems from the Protocol."

A further consequence "is the increased likelihood of proliferation of chemical weapons and the breakdown of the restraints against chemical warfare," CBW expert Matthew Meselson warned in a 1970 article. "Our employment of harassing gas in war, particularly when it is done on a large scale in conjunction with ordinary military operations, stimulates military planners in other nations to [organize their own CW programs]," he added.

Meselson also noted chemical herbicides' potential for causing ecological and public health catastrophes—since borne out grotesquely in the victims of Agent Orange. "Because of the relative ease with which this tactic can be practiced," he wrote, "it would be difficult to stop once the precedent is set."

Despite gaping *legal* loopholes, the treaty provides useful *political* barriers. All CBW outlaws—regardless of treaty status—risk diplomatic sanctions, the threat of shifting alliances, and the loss of international prestige. American influence in world arms control policy took a tremendous international beating over the failure to ratify the Geneva Protocol and use of CS and Agent Orange in Vietnam. Ultimately this pressure affected U.S. behavior and, to a degree, U.S. policy.

United Nations resolutions on CBW, while legally toothless, are powerful weapons for the propaganda machines of East or West. For poor nations dependent on international aid, avoidance of pariah status is a potent motivator. Trade restrictions and other economic measures can be equally influential. (As shown in the next chapter, however, international legal restrictions may have an unwanted by-product: Potential violators may increase the level of secrecy in their already highly classified CBW programs, shielding them further from public scrutiny and from assessments of the potential for violations.)

Perhaps the most powerful effect of the protocol is *moral*. It codifies and reinforces worldwide aversions to CBW that led to the treaty in the first place. "Use of such weapons has been outlawed by the general opinion of civilized mankind," President Roosevelt stated in 1943. The protocol serves as an explicit reminder of this worldview—regardless of its legal power.

WHY TREATIES ARE VIOLATED

During World War I nations learned through bitter experience that gas warfare between armies that possess defensive equipment is tactically futile. It increases casualties and slows the course of battle painfully, but it fails to provide a decisive advantage. This lesson has helped to discourage the use of CW ever since. But the moral and political suasion of the Geneva Protocol has also played a compelling role. Although accusations have been legion, even under the most one-sided circumstances only the six party nations indicated in Table 3 have verifiably violated the protocol since its inception.

TABLE 3

VIOLATIONS OF THE GENEVA PROTOCOL*
(INCLUDING TEAR GAS AND HERBICIDES†)

VIOLATOR/TARGET	YEAR	WEAPON USED
Egypt/Yemen	1963–67	CW of unverified type
France/North Africa	Late 1950s	CW/herbicides
Great Britain/Malaya	Early 1950s	CW/herbicides
Iraq/Iran	1980s	CW/various
Italy/Abyssinia (Ethiopia)	1936	CW/mustard gas
Portugal/colonies (Angola, Guinea-Bissau, Mozambique	1968–74	CW/herbicides

* Japan conducted extensive CBW attacks on the Chinese from 1937 to 1942 but was not a party to the protocol until 1970.
† The United States was not a party to the protocol during its period of herbicide and CS use in Indochina.

Sources: SIPRI, Chinese, Japanese, and U.S. government sources.
See source notes for details.

At least eleven other nations have been accused of violating the protocol. And many nonparties may have used CBW, while dozens of others have undoubtedly considered CBW attacks, regardless of the consequences. The foremost rationalization is desperation. As shown by the Iraqis, fear of annihilation may override any kind of restriction. And as noted in Chapter 2, even major powers succumb to such fears. The British, in particular, laid plans for massive gas attacks on Nazi Germany in the face of devastating V-2 rocket raids.

The perception that CBW would provide a decisive advantage has at times been an irresistible incentive. Italy violated the protocol on a massive scale and in cold blood, scoring overwhelming military successes against the pitifully weak Ethiopians. The Japanese attacks against China did not violate any law (see Table 3), but their rationale was the same as the Italians'.

Similarly, the conviction, however misguided, that CBW could save lives by shortening a conflict has been a contributing factor, as reflected in Winston Churchill's previously mentioned consideration of CBW during World War II.

An equally influential factor in treaty violations is the expectation that an enemy is preparing to "break out" first. In 1939, following Hitler's invasion of Poland, British Chief of Air Staff Sir Cyril Newall told his colleagues that consideration was being given to the "deliberate and indiscriminate dropping of bacteria with the object of spreading disease." Newall's stated rationale: "The fact that the German Government has notified us of their intention to observe the Geneva Protocol is, of course, no reason to imagine that they will in fact observe those provisions a moment longer than is necessary."

As in other arms agreements, new technologies play no small part in decisions to use CBW in violation of the law. Italy's mustard gas attack on Ethiopia in 1936 was based on the devastating effectiveness of the world's first aerial spraying of poison gas.

By the same token, advances in weapons technology—legally permitted—can exert pressure to redefine the protocol's meaning. A British policy switch on harassing gases proves this point. When it signed the protocol in 1925, Britain considered the gas provisions all-encompassing. But in 1970 this interpretation changed. Britain began to view the protocol as banning only the use of harassing gases known in 1925. CS was not weaponized until after World War II, therefore was legal for use under the 1970 policy formulation.

The official justification was that CS is less toxic and therefore more humane than earlier tear gases. The distinction may have been drawn, however, to blunt criticism of the British Army's use of CS for riot control in Northern Ireland.

Secrecy and deniability are among the most important factors in consideration of treaty violations, as so well exemplified by the CIA toxin scandal of 1975 and the allegations of agency sabotage of Cuban agriculture with BW agents. Spy agencies know that secrecy or deniability is essential not only if the activity is a violation of international law but also if socially and politically reprehensible acts are to be effective as tools of policy.

The toxin affair was scandalous because it was made public. In contrast, because the Cuban allegations are still a matter for speculation and unsupported by conclusive evidence, they have triggered negligible international repercussions. By the same token, despite its ultimate debunking, the case for yellow rain provoked considerable initial outcry because physical evidence was offered to substantiate the accusations. In this way denials were initially suspect.

NEGOTIATIONS IN CW DISARMAMENT

Multilateral talks to bolster the CBW arms control regime began in 1968 at the thirty-one-nation Conference of the Committee on Disarmament. In 1978 the name was changed to Committee on Disarmament and the body was expanded to forty nations, including most WTO and NATO countries, as well as major nonaligned states. The groundbreaking 1972 BW and toxin ban, the committee's proudest CBW achievement, is discussed in detail in Chapter 7. Efforts have also been made to negotiate a CW disarmament treaty, but progress has been sluggish.

Bilateral talks between the Americans and Soviets have also stumbled along for years without substantive gain. Periodically, fresh initiatives are floated. More frequently propaganda wars are launched when it suits one side or the other to illustrate a particular bargaining position or to gain domestic support for a new weapons system. A new CW convention that bans *possession* of all chemical arms remains an elusive goal.

The U.S.-Soviet talks have been marked by heated debate over various issues, such as the particular chemicals and production facilities that should be included in a treaty. The talks have foundered, however, on general concern over verification. The problem is highly complex. Unlike intercontinental missiles, chemicals can be produced and stored in a multitude of small facilities, thus potentially obscuring offensive development.

For years the United States has pressed for Soviet agreement to nearly unlimited on-site verification on demand. The Soviets have consistently demurred, fearing that inspectors may gain unwarranted access to classified defense information and that potential CW sites may be defined unfairly. For example, would the United States allow fully reciprocal site visits by Soviet inspectors to commercial plants capable of producing chemical weapons?

But more than just technical verification differences have mired the talks for nearly two decades. Each superpower fears the other's new technologies. Each accuses the other of planning a technological breakout that will render existing arsenals obsolete. American development of binary nerve gas has evoked just such concerns from the Soviets. Because of their low toxicity in storage and handling, the Soviets fear that binaries could assume an increasingly important and flexible role in NATO planning once they are deployed. This would make them difficult to dislodge, even for the sake of a disarmament treaty.

"The more important the weapons come to appear," Julian Perry Robinson of the University of Sussex notes, "the more will seem to be at stake in agreeing to forswear them."

The prospects for dramatic Soviet modernization are less predictable. The United States freely admits that its intelligence gathering on CBW has been lax since 1969. The lack of effective threat assessment confounds efforts to form a coherent defensive strategy or to appraise prospects for a *Soviet* technological breakout. Where does that leave the United States? Fearful and suspicious.

After weapons innovations have been around for a while, they are usually accepted regardless of initial qualms. For example, the 1868 Declaration of Petersburg—still technically in force—outlaws certain projectiles that are "either explosive or charged with fulminating or inflammable substances"—in other words, hand grenades, bombs, missiles, and torpedoes. As technology grows more sophisticated, moral and legal standards tend to shift in accommodation. The new biotechnologies are exerting the same kind of acceptance pressure on the body of international law that forbids the use of CBW.

By the same token, treaties may be eroded or rendered obsolete or irrelevant by technological advances. The rules governing the passage of ships in disputed waters are a case in point. They mandate a warning shot across the bow of an enemy vessel. But how do you fire such a shot against a ship that is far beyond visual range and if your only weapons are radar-guided missiles that cannot be made to miss? The answer, as dramatically shown in U.S. encounters with Libya in the Gulf of Sidra in 1985, is: You can't. Instead, you simply sink the enemy ship.

Proliferation adds massive new complications to chemical disarmament efforts. It was difficult enough to disarm when only three major powers had CW stockpiles. As noted in the last chapter, the CW club is growing rapidly.

The nuclear question is never far from the minds of CW negotiators. No nuclear power has ever used CBW against another. Some experts argue that large-scale use of CW would inevitably lead to a nuclear exchange; others are equally convinced that chemicals are our best hope to stop a conflict from escalating into nuclear holocaust. This rhetorical tango has, of course, only contributed to the inertia on CW disarmament.

Likewise, difficulties the superpowers have encountered since 1979 in reaching new agreements on nuclear weapons may have confounded progress on CW. "The U.S., having pressed hard for on-site inspection in the Comprehensive Test Ban talks, may therefore be less likely to accept a more lenient regime in the CW talks, for fear of setting a precedent," observed Canadian defense analyst G. K. Vachon. "This, combined with a superpower preoccupation with nuclear weapons, sets the scene for a deadlock in negotiations on chemical weapons."

U.S.-USSR eagerness for—and progress toward—a closely monitored medium-range missile agreement in 1987, however, appeared to stimulate at last some hopeful signs on the chemical negotiations front. In April of that year Soviet bloc and American representatives to the Committee on Disarmament began to express cautious optimism that a CW ban could be signed within a year. They indicated confidence that the sticky verification issues will eventually be resolved. And in October 1987, the Soviets displayed a CW center to diplomats and experts from the United States and forty-four other countries for the first time.

Superpower hegemony over the negotiations process, however, has exacerbated chemical disarmament dilemmas and contributes to the skepticism with which many analysts view apparent diplomatic progress. Bilateral talks between the Soviets and Americans are vital to CBW arms control, of course. The problem is, they have usually not contributed to CW treaty prospects—in part because of the complex agendas brought to the table. At the same time bilateral talks limit the role of other nations, such as those of Central Europe, whose security interests in rapid progress are more urgent. In most scenarios these countries stand to bear the brunt of a biochemical conflict between the superpowers.

In the past few years, however, growing efforts by the opposition Social Democratic party in West Germany and by East German officials have been under way to create a chemical-free zone in Central Europe. This is a corollary to the chemical-free Europe plan proposed repeatedly by the Soviet Union. The Germans envision an eventual extension of the limits to a comprehensive ban.

The United States and NATO reject regional plans out of fear that the Soviets would more easily be able to supply troops in Europe with CW—an open invitation to a first strike. The plan has also been called a device to divert interest in negotiating for global disarmament.

THE PSYCHOLOGICAL DIMENSION

Vexing as they are, these technical questions pale beside a profound crisis of confidence. U.S. allegations of Soviet violations of the Geneva Protocol and other treaties have sounded a steady drumbeat since 1981. But firm evidence, sufficient to convince either arms control experts or U.S. allies, has not emerged. Some influential U.S. officials clearly oppose all arms control treaties because they believe the Soviets will never adhere to their provisions. (This issue is discussed in detail in Chapter 10.) The Soviets, they say, are on a single-minded quest for total military superiority.

"I don't believe that the argument that you can't trust the Russians so you shouldn't have treaties with them has any merit at all," said Matthew Meselson in 1984. "Not because you can trust the Russians, but because treaties should never be based on trust. Treaties should be based, when needed, on verifiable provisions, but these allegations have muddied the waters and hang as a miasma over the negotiation process."

This pattern of behavior has heightened concern about whether the Reagan administration is genuinely interested in a CW ban. Jorma Miettinen, an arms control expert at the University of Helsinki, had this to say about the twenty-four-hour "unimpeded access" verification provisions of a much-heralded U.S. proposal carried personally to Geneva by Vice President George Bush in 1984: "Even the allies of the United States feel that it is written such that no treaty would be signed."

On three occasions the vice president's votes in favor of binary weapons broke Senate deadlocks. Thus the choice of Bush to present the proposal, commented syndicated columnist Mary McGrory, "is also in the great Reagan tradition of picking a negotiator whose record suggests that he is only kidding as he opens his samples case."

For their part the Soviets have done little to enhance their own credibility. They have rarely responded substantively to detailed charges, such as in the case of yellow rain. Instead, they have flatly rejected U.S. accusations as disinformation. At times they have weighed in with farfetched, unsupported countercharges—such as the suggestion that AIDS may be the product of an American BW experiment gone awry. And their CBW program is shrouded in such secrecy as to make the United States, with its limited revelations, look like a paragon of openness.

These tensions have poisoned the CW disarmament process. As will be shown in later chapters, they now gravely threaten biological arms control.

5

A REVOLUTION IN BIOLOGY AND WEAPONRY

RECONSIDERING CONVENTIONAL WISDOM

When rDNA and other new biotechnologies burst onto scientific center stage during the 1970s and early 1980s, their weapons implications caused an inevitable reevaluation. Long-standing doctrines for biological warfare were suddenly called into question. Presented below as a basis for analyzing how BW policy has changed are the criteria for an "ideal" antipersonnel BW agent that generally have been accepted for decades. Adapted slightly, many of the same factors apply to antianimal and anti-plant agents:

1. Highly virulent, consistently causing disease or death at low concentrations and infective through multiple routes, such as respiration, ingestion, and skin contact.
2. Highly contagious—if possible transmissible from specific animals to

humans but with limited general spillover that would create an uncontrollable epidemic among many species.

3. Short, predictable incubation period from exposure to onset of symptoms.
4. Robust in the environment under adverse conditions, but short-lived, with predictable persistence if target area is to be occupied by attacking troops.
5. Deniable—either endemic to the target region or causing symptoms that mimic those of endemic diseases.
6. Medical defense difficult—target population or army should have little or no natural or acquired immunity and little recourse to immediate, effective prophylaxis or medical treatment.
7. Countermeasures ultimately available—vaccines or other effective protection, as well as decontamination methods, should be available to aggressor troops, civilians, and ultimately target groups after occupation.
8. Suitable for economical mass production, stable within munitions during storage and transportation, and amenable to dissemination by aerosol spraying, explosion, or living vectors (such as mosquitoes or fleas).
9. Psychologically effective—the attack should produce terror or demoralization.

An additional criterion would be added for antianimal and antiplant BW agent:

10. As host-specific as possible to reduce the risk of generalized damage to the biosphere.

An example suggested by Rand Corporation analyst Raymond Zilinskas shows how an agent with these attributes might work:

Rabbits are not native to Australia. After they were carelessly introduced there in 1859, their characteristic explosive proliferation soon spawned a major pest control problem. By the 1940s the voracious creatures were driving scores of farmers to bankruptcy. The Australians attempted many methods of control, including the erection of a 1,100-mile-long "rabbitproof" fence. None of these was successful for long. The rabbits were ultimately defeated, however, through infection by myxoma virus, causative agent for the disease myxomatosis.

Myxoma is a model BW agent. The virus is harmless to all species except the rabbit. And for the type of rabbit targeted in Australia it was highly virulent, causing 90 percent fatalities and sterility in many of the survivors. "For the purpose of decimating rabbit populations, the agent is stable and easily dispensed," Zilinskas said. Although an effective vaccine exists for the disease, the rabbits, of course, had no access to it.

Myxomatosis "may be partially controlled by seasonal use and by a thorough consideration of geographical and environmental factors"—an acceptable level of control considering the high host specificity. Of course, the element of camouflage—using an organism that is either endemic to the region or resembles endemic diseases—is irrelevant to pest control. But a myxoma epidemic could easily have begun naturally in Australia.

Despite decades of research, not a single BW agent, including those standardized as weapons by the United States, has come as close as myxoma virus to the BW ideal. Anthrax and bubonic plague, for example, are highly virulent, stable, and hardy enough for effective dissemination. Each attacks rapaciously and causes horrifying symptoms. Anthrax, for example, causes large, black skin ulcers. Although these diseases can be identified and treated effectively with antibiotics, many people would die from an attack before medical countermeasures could be put into place.

But these agents are highly persistent in the environment. Plague may easily form a reservoir in animal populations and can spread over wide areas via rodent and flea vectors. Anthrax remains dormant but potentially infectious in spore form for decades. The long- and short-range potential for accidental infection of aggressors or their allies in close geographic proximity to an attack is extremely high. And a major epidemic of anthrax or plague would be a rare occurrence almost anywhere in the world. Because natural causes would be improbable, political and health officials would immediately scrutinize the epidemic intensively.

Building and using an effective, reliable germ weapon are a daunting and dangerous technical undertaking. Like anthrax and plague, many agents fulfill a few key weaponization characteristics, but few boast a combination sufficient to convince skeptical field commanders, who want weapons they can depend on. Because biological weapons are essentially uncontrollable and act unpredictably in the environment and within their victims, they have rarely been used in war, and never to decisive advantage in modern times.

This is not to suggest that conventional biological warfare is totally unfeasible. Plants, which lack an immune system, have always been particularly susceptible to biological agents. As we've seen, monocultivation—the standard in world agriculture—offers ideal BW targets.

Massive biological crop destruction, timed with the growing season to exclude effective replanting, could reap a devastating harvest.

And as the examples in earlier chapters show, under many circumstances antipersonnel BW attacks, while imperfect, can be successful. A government may be tempted to use BW whenever the depth of the crisis appears to outweigh the pitfalls. But understandably, generals and admirals have always sought something more like myxoma virus—controllable, specific, deadly.

THE OPTIONS BROADEN

For the first time military planners now believe these ideal BW agents are on the horizon. Genetic engineering, they say, revolutionizes the potential of every aspect of BW, quantitatively and qualitatively. It offers the prospect of dream weapons—vastly more effective than ever before. These are the primary ways novel BW are envisioned:

Increased Pathogenicity

Nature has produced a wide array of frightfully dangerous organisms. But altering specific agents to increase their virulence could greatly enhance efficiency, while preserving deniability in a biological attack. A particularly powerful strain of an endemic pathogen could simply be blamed on a chance natural mutation.

Although some of the following examples apply equally to bacteria and rickettsia, viruses are considered the most efficacious BW agents. Because different viruses often cause similar symptoms, viral diseases are often difficult to diagnose. And viruses are unaffected by a broad range of drug therapies that are lethal to infecting bacteria.

Although virulence is poorly understood, molecular biologists suggest several methods that could create more deadly viruses. For example, many viral diseases are caused by toxins produced on the instruction of one or more identifiable tox-genes. The virus injects these tox-genes, along with the rest of its genetic material, into a host cell. The host then effectively destroys itself by manufacturing the viral toxin coded by the tox-genes. Manipulation of tox-genes could result in the production of greater quantities of more pathogenic toxins.

Viruses could also be used as effective vectors for one or more powerful tox-genes implanted through rDNA techniques. Other genes control the ability of the viral surface to identify specific receptor sites on the surfaces of cells, which provide points of entry in a viral attack. Mutating or rearranging these genes could bestow the ability to recognize more attack sites, producing a more invasive organism.

"The immune system is also damaged by irradiation as well as numerous toxins and poisons, including CW and TW agents," according to East German BW expert Erhard Geissler. "Thus additive and even synergistic effects result if humans more or less damaged [by these other weapons] are infected by BW agents."

Defeating Immunity and Diagnostic Testing

Acquired immune deficiency syndrome (AIDS) demonstrates an even more frightening function of some viruses: paralysis of the immune system. Virulence is thus indirect; other pathogens—such as a mild flu—normally overcome by naturally produced antibodies suddenly become killers. The Soviets have called AIDS a U.S. BW experiment gone out of control. Although no evidence has been presented to support the claim, manipulating genes to defeat the body's immune response is quite feasible and consistent with some U.S. studies.

Immune defense (see Chapter 1), whether natural or acquired through vaccination, is based on antibodies targeting precise surface sites (the "antigenic structure") on the invading virus or bacterium. The genes that encode such antigenic surfaces have already been identified for many disease-causing organisms. Genetic manipulation aimed at altering the antigenic structure of BW agents could render defending antibodies "blind" to their targets. This type of genetic mutation occurs regularly in nature. People catch the flu year after year because slight changes in the flu virus surface can render ineffective resistance developed for the previous year's strain.

Recently monoclonal antibodies have been developed as highly sensitive diagnostic tools, as described in Chapter 1. Like the immune system, MCAs rely on correct identification of specific sites on the BW agent's antigenic structure. Thus, altered organisms that could defeat the body's natural defenses would also neutralize these important diagnostic reagents.

Controllability

Genetic engineering has opened several avenues for approaching the most vexing problem of BW design: how to stop a weapon before it overshoots its intended goal or permanently implants itself in the ecosystem.

"The attacker may be able to retain control over the BW agent by designing it to: (a) cause a maximum effect on the host in a short time . . . ; (b) to die off after a previously determined number of cell divisions; (c) not cause any secondary infections; or (d) by designing the organism to be bound by a narrow set of environmental factors. For example, if the temperature falls under 15 degrees Centigrade the organisms would be unable to survive," Zilinskas has speculated.

To expand the point, he suggests weaponizing the innocuous laboratory workhorse *E. coli,* an organism that typically lives in the human gut, by implanting the gene for botulinum toxin, one of the deadliest poisons. "Botulin-producing *E. coli* (*E. Coli* Bot+) could be liberally added to water supplies or various food preparations in order to reach the widest scope of targets. Most probably, the altered *E. coli* would not be able to survive competition from wild *E. coli,* thereby insuring the elimination of *E coli* Bot+ after a few generations and also after it has been given a chance to eliminate large numbers of people."

This is one example among many. Advances in molecular biology since 1981 have made this kind of manipulation straightforward.

Drug Resistance

Like all characteristics of cells and viruses, bacterial resistance to antibiotics and viral resistance to other drugs are determined by specific genes. Drug resistance genes are often transferred between different organisms in nature. With rDNA technology it is a simple matter to mimic this process, an obvious offensive enhancement for a germ weapon.

This is a sobering prospect, as illustrated in this scenario suggested by Zilinskas:

> Nation A is at war with B. A knows about the usual medical practices in B, including the accepted means employed to treat dysentery. In B the first choice antibiotic . . . is ampicillin, the second is streptomycin, and the third is chloramphenicol. A also knows that the facilities for producing cephalasporins [are] poor or nonexistent in B. A now produces large quantities of *Shigella* resistant to the first three antibiotics but sensitive to the fourth. The resistant *Shigella* is thereupon introduced into B's water and food to the fullest extent possible, while, at the same time, A makes certain its own military and civilian medical services have a full supply of cephalosporins on hand.

Increased Environmental Survivability of BW Agents

Aerosolization is the most efficient method for disseminating a biological agent. The problem is that most BW agents have evolved for survival inside living organisms; only a few, such as anthrax, are hardy enough to withstand atmospheric challenge. Microencapsulation has been applied with increasing success to this problem. Microencapsulation is a method of enclosing microscopic drops or particles within individual protective sheaths composed of certain organic compounds. The technique, introduced in 1954, is used for everything from carbonless copy forms to timed-release pharmaceuticals. Aerosolized agents may soon be protected from the sun's radiation, immediate drying, widely varying temperatures, the explosive force of munitions, even the stress of sprayings from jet aircraft traveling at supersonic speeds.

In addition to microencapsulation, rDNA could be used to "toughen" a BW agent. The genes known to control cellular repair could be manipulated to enhance the survival of an agent faced by these kinds of environmental stresses.

Along these lines falls ongoing U.S. military research into arthropod vectors. After all, nature's own method of transferring some of the most disastrous diseases, such as malaria or dengue, is to enclose them within the mosquito. As understanding of the relationship between pathogens and arthropods grows, "more efficient (or even totally new) host-vector systems" should be possible, Geissler has pointed out.

Ethnic Weapons

As members of a single species all humans have similar genetic blueprints. But variations in genes occur across particular populations or racial groups. Some of these genetic differences are dramatic. For example, to about 70 percent of all people, phenylthiourea tastes bitter. Yet barely 1 percent of certain groups of Brazilian aboriginals can taste the chemical.

"An enzyme deficiency in Southeastern Asian populations [makes] them susceptible to a poison to which Caucasoids are largely adapted," Swedish geneticist Carl Larson wrote in 1970. "The poison at issue is milk."

Larson's seminal article, published in the U.S. Defense Department publication *Military Review,* stirred arms control experts because he suggested that these differences in genetic predisposition might be exploited to create genotype-targeted CBW—that is, weapons that discriminate on the basis of race or ethnicity. This was the first time the subject was broached publicly. But in the military's private circles it was old news.

In 1951 the U.S. Navy conducted a series of top secret tests on its Mechanicsburg, Pennsylvania, supply depot, involving a BW simulant, a benign organism used to mimic the behavior of an actual weapon. According to documents declassified in the late 1970s, the site was chosen because "Within this system there are employed large numbers of laborers, including many Negroes, whose incapacitation would seriously affect the operation of the supply system. Since Negroes are more susceptible to *Coccidioides* than are whites, this fungus was simulated by using *Aspergillus fumigatus*."

Coccidioides immitis is a fungus that causes coccidioidomycosis, commonly called valley fever, after California's San Joaquin Valley, where it is endemic. Valley fever is a systemic, sometimes fatal disease that the military has studied as a potential BW agent from the 1940s to the present. Blacks are up to ten times as likely to die from the disease as whites.

Advances in biotechnology and human genetics have made the prospect of identifying these discriminating agents far easier than ever before. A recent discovery in tumor virology is a case in point. "Epstein-Barr virus . . . exerts quite different pathological syndromes in different populations," notes a 1984 SIPRI report. The virus gives whites infectious mononucleosis, a benign disease, while black Africans and Southeast Asians develop two different forms of cancer.

On its face, speculation about such genocidal weapons may seem too bizarre to take seriously. After all, although scientists have discovered at least twelve genes that affect sensitivity to particular toxic substances, not one of these will split any two population groups cleanly. The "perfect" ethnic weapon—perhaps sought by a white supremacist regime to kill any black person while sparing all Caucasians—is probably impossible.

Perfection, however, has little to do with military expedience. Even if the difference in susceptibility between the aggressor and target populations were only 90 percent, or even 60 percent, the advantages—including the obvious mass psychological impact—may be perceived to outweigh the potential risks. And even these risks can be minimized through careful "target analysis and the selection of personnel for special missions," Larson pointed out.

Biochemical Weapons

In 1985 a U.S. Army report noted:

> It is possible to artificially produce the natural biological substances which exert potent regulatory effects on the body. These substances are normally present in the body in minute amounts and control mental states; mood and emotion; perception; organ function, growth and repair; temperature; and other body processes. These substances are not considered to be toxic and are indispensable to the normal functioning of the human body. But even slight imbalances can cause profound physiological effects, leading to incapacitation and even death.

Of course, the hormones, hormone receptors, and cellular "growth factors" (similar to hormones) the army was referring to have been known for years, but until the advent of rDNA obtaining even minute quantities was laborious and prohibitively expensive. Now all these substances can be produced in large amounts by molecular cloning. In fact, in this form they are stable, simple to store, and rapidly expandable.

Novel Toxin Development and Manufacturing Methods

Until the late 1970s large, complex protein molecules—such as interferon (the immunity enhancement drug), growth hormones, and toxins—could not be synthesized chemically. Instead, they were extracted and purified from animal tissue or plants. In view of these constraints, it was impractical to produce anything but bacterial toxins—such as botulinum—in sufficient quantities for military stockpiling. Toxin weapons were primarily considered the tools of spies—useful for covert assassinations.

Times have changed. Through molecular cloning toxins now can be synthesized cheaply and in unlimited quantities. (Genes that code for many of the most potent toxins, including botulinum, diphtheria, anthrax, and cholera have already been cloned by the U.S. Army.) Because of their extreme toxicity (some far more toxic than nerve gas by weight), if they could be made to act rapidly, toxin weapons could theoretically be used effectively in much smaller volumes than their chemical counterparts.

Of course, bioengineering has also generated revolutionary insights into the basic biochemical structure and function of all living systems.

"The target structures [such as cell membranes and toxin-degrading enzymes] of both conventional and new CBW agents can be identified and described in molecular terms, which would enable CBW agents to be tailored more precisely," according to a 1984 analysis by SIPRI. "In addition, those structures and functions which interfere with the toxic action of any given CBW agent, including its penetration and its stability and half-life in an attacked host organism, can be identified and characterized. This too would be useful in tailoring more efficient CBW agents."

For example, scientists know nerve gases and some neurotoxins work by inhibiting the enzyme acetylcholinesterase (see Chapter 2). But the precise mode of action is still a mystery. Using genetic engineering, military and civilian labs around the country are working to clone the gene that codes for this enzyme, in order to produce a large quantity of acetylcholinesterase for experimentation. Such work could elucidate the precise mode of action of these weapons and generate knowledge leading to agents far more potent than those currently available.

BIO COUNTERMEASURES

The standard medical countermeasure to nerve gas is automatic injectors using the chemicals atropine and 2-PAM (pralidoximine) chloride. If they are used immediately after exposure, "the chances of recovery are reasonable, according to Colonel Rudolph de Jong, of the U.S. army's Medical Research Institute of Chemical Defense.

Unfortunately immediate use may be a vain hope in the heat of battle. Many soldiers would die before the antidote made it out of their backpacks. And the drugs are of limited effectiveness. Other antidotes must be administered promptly in a field hospital. Atropine also has serious side effects if it is used erroneously during the inevitable false alarms of any ongoing chemical attack. The biotechnology research described above is already raising hopes that more effective antidotes for acetylcholinesterase inhibitors will soon be available.

Studies that could increase pathogenicity or defeat the immune system can also improve prophylaxis by leading to vaccines for BW agents and to toxoids and antitoxins for TW agents. Vaccines are: noninfectious strains of normally pathogenic bacteria and viruses, pathogens that have been killed or inactivated with chemicals or radiation, or fragments split off from pathogens (see Chapter 1). Toxoids induce immunity to toxins in the way killed vaccines induce immunity to living organisms. Antitoxins are analgous to antibodies that target toxins.

Effective vaccines or toxoids owe their success to their specificity. They stimulate the immune system to recognize, attack, and destroy a particular antigenic signal; for example, anthrax vaccine protects against anthrax but is useless against plague or cholera. Many leading scientists now believe that new biotechnologies can develop effective prophylaxis for any individual disease or toxin. Indeed, viral vaccine development is the largest stated goal of the DOD biotechnology program.

The trouble is, in a defensive posture it is difficult to know which vaccinations are needed. A dengue vaccine could be distributed, but what if the enemy attacked with Rift Valley fever? Not only are multiple vaccinations impractical and expensive, but many vaccines cause negative side effects and may be effective for only a few months or years. Potentially dangerous, continual doses of vaccines would be required for soldiers at high risk.

Vaccinia virus, which eradicated smallpox as a significant health hazard, raises new possibilities. Vaccinia is a relatively large virus with an unusual surface. Researchers have been able to splice genes coding for the surface coats of other viruses, such as influenza, hepatitis, and rabies, into vaccinia virus DNA. The result: a vaccine with a coat of many colors—a "broad spectrum" vaccine, conferring immunity for up to three diseases at a time in both experimental animals and humans.

However, a recombinant vaccinia that protects against dozens of diseases is not on the immediate horizon. "Whether the immune system could respond adequately to such an intrusion remains to be established," one expert in the field notes. But work to create the ultimate vaccine has been promising beyond the wildest fantasies of a decade ago.

Similarly, striking advances in MCA technology suggest that while new vaccines are being developed for long-term immunity, antibody injections conferring temporary immunity can be prepared for a myriad of BW and TW agents. Some speculation—and considerable military funding—are even aimed at creating vaccinelike pretreatments against chemical weapons. These would target the organophosphorus poisons that are the active ingredient of nerve agents.

Another fruitful area is antiviral drugs. Alone or in combination with such immunity enhancers as interferon, drugs like ribavirin can retard the growth of a range of viruses. Military scientists believe this work could lead to reliable treatments against entire classes of viruses. Such a development—like antibiotics, which kill bacteria—would obviously be a monumental medical achievement.

Researchers have also discovered that microencapsulation is good for more than dissemination of CBW agents. They hope the technique can be used to direct drugs such as ribavirin to particular sites in the body. In recent years efficient methods have been devised for producing liposomes, tiny spheres composed of materials similar to cell membranes, that can enclose, stabilize, and carry virtually anything—therapeutic drugs, toxins, viruses, even rDNA. The liposomes can be administered by a variety of routes, including aerosols, and they are quite stable once within the body. Furthermore, by the coupling of such molecules as monoclonal antibodies to the liposome coat, the microcapsules can be selectively "targeted" to cells that display the corresponding antigen. For lung infections, for example, these capsules could be focused directly to the lungs rather than be diluted throughout the body.

OTHER BIOTECHNOLOGICAL FACTORS

Several corollary developments in the biological sciences also have potentially profound influence on CBW.

Laboratory Safety and Production Standards

Working on the weaponization of deadly diseases, many of which have no known cure, will never be child's play. During the heyday of the U.S. BW program Fort Detrick researchers posted an infection rate nearly three times that of their counterparts at the National Institutes of Health and more than seven times that of those at the National Communicable Disease Center. The work was risky business, in part because relatively large volumes of the deadly organism had to be grown in order to isolate and test its subparts. A false move at any step in a complex process could be fatal.

But rDNA offers a dramatic safety improvement. Pathogens can now be studied in the relative safety of "biological containment," in which their genetic material is reproduced, examined, and manipulated within special strains of *E. coli* or other host bacteria that have been crippled so that they cannot survive outside a narrow range of nurturing experimental conditions. This greatly reduces the likelihood of an organism's infecting a researcher or escaping from the lab.

The rDNA guidelines established in the 1970s (discussed in detail in Chapter 8) also standardized and codified important new physical containment conditions for working with hazardous biological materials.

Bioprocess Improvements

Until the early 1980s the production of militarily significant quantities of both biological weapons and vaccines required industrial-scale facilities. Fermentors, storage tanks, and their manufacturing infrastructure were massive, dangerous to operate, and difficult to maintain, conceal, and protect. New bioprocess technologies have closed the book on these problems.

To be sure, growing, handling, storing, and transporting BW agents will always be extremely dangerous. But new techniques have wrought sweeping changes in production efficiency and efficacy. The following section of a U.S. Army report to the House of Representatives shows how seriously the military takes these developments:

> Recent advances . . . make possible the growth of mammalian cells on the surface of minute beads, rather than the inner surface of glass roller bottles. . . . One small bottle partially filled with beads can now be used to produce virus yields which previously required large production facilities operating under stringent conditions. This new technique greatly simplifies virus production and allows large-scale yields in facilities of very modest size which can easily be hidden.
>
> The introduction of continuous flow fermentors also increases productivity. Most likely the size of fermentors operating by batch process [the classical method] can be reduced a thousandfold by conversion to a continuous flow process. . . . [Soon] the major limiting factor will be storage space. . . . In some cases storage will not be needed because large quantities can be produced very quickly from clandestine sources. For example, producing one pound of antibodies traditionally required 50,000 mice or 22,000 gallons of tissue culture media. The use of encapsulated hybridomas lowers the total volume requirement to 220 gallons—a hundredfold reduction. . . .
>
> Hollow fiber technology permits the fifteenfold concentration of influenza virus in 14 hours with 70–90 percent recovery versus 30 hours with 40–50 percent recovery using bulkier conventional [methods].

Finally, the report notes, the time needed to isolate pathogens from other biological materials has been reduced by a factor of 100.

Ultrasensors

You're a soldier for nation X. Enemy planes from nation Y are flying bombing runs overhead, and the air is hazy from explosions. Your infantry unit could be called to move through treacherous battle zones to a strategic bridge at any time. Because Y has used CBW in the past, you're carrying a state-of-the-art protective suit. But the heavy, hot, bulky gear would slow down your movement dramatically, making it a last resort. In addition, nearly all modern CBW agents are colorless, odorless, silent killers. How will you know when to put on the lifesaving suit?

According to CBW researchers, you'll know because your unit's ultrasensor will register an alarm when a deadly chemical, toxin, or biological organism hits it—before concentrations reach harmful levels. This is the idea, anyway, behind portable sensors, now under development, which could identify dangerous organisms encountered in routine or battlefield conditions with exquisite specificity. The most promising biological sensors use monoclonal antibodies that detect minute amounts of a particular antigen. Theoretically these devices would be carried with troops and posted at the periphery of encampments or battle zones.

Aerosol Immunization

Development of new vaccines and their effective mass production are one thing. At the point of mass inoculation, a problem arises: How can millions of people be immunized quickly and without causing panic?

The human nose and air passages, first lines of defense against infection, may provide part of the answer. "There is the sticky substance called mucus that traps many particles. Farther inside are threadlike cilia that cover some of the surface tissues like beds of waving grass, sweeping out any particles big enough to catch," in the words of one science writer. "The pink, velvety mucous membranes that line the airways have other potent defenses, too. Protective antibodies lurk in their surface layers. Scavenger cells cruise the territory to engulf invaders and destroy them."

These methods are not always successful at foiling the entry of disease organisms. This, of course, is the reason for vaccines: to stimulate immunity when spontaneous systems are inadequate. Normally vaccines are injected or taken orally. Aerosol vaccines—inhaled through the nose and mouth—have been explored by the medical profession but are considered of limited usefulness because their dosage is difficult to regulate and their ability to confer immunity is unproved.

Most naturally occurring epidemics are spread by insects or other biological vectors or directly from person to person. But the most probable method of dissemination for many BW agents is an aerosol. This is the reason, the DOD says, for its intensive study of aerosol immunization. According to one army report, this work is targeted toward "stimulating protective immunity on mucosal surfaces throughout the respiratory tract."

U.S. troops are routinely inoculated for a variety of diseases, and large-scale civilian vaccination is certainly nothing new. But mass vaccination against rare or exotic diseases, which include many of the most likely BW candidates, would touch off a public furor regardless of the stated justification for it. This may be another reason for military interest in aerosols. Theoretically, spraying a vaccine over wide areas could covertly inoculate large populations.

Although this idea may seem like the stuff of science fiction, the army has long taken it very seriously. As early as 1963 an article in the journal *Military Medicine* noted that "a plan for large-scale immunoprophylaxis of the civilian population should be prepared. This would include standby legislation for compulsory immunization." A separate article cites aerosol vaccination as a means to accomplish that goal. In today's world, of course, this method would more likely be used covertly to vaccinate allied troops or civilians in the third world to protect them against secret BW attacks on nearby enemies.

And the massive, secret dissemination of biological organisms during army tests in the 1950s—over San Francisco Bay and through the New York subway system, among other places (see Chapter 2)—unquestionably verifies that aerosol vaccines can be covertly disseminated.

Predicting the Spread of Disease

For decades epidemiologists have sought a framework to predict what have been seen as fundamentally unpredictable events—the spread of epidemics—in order to limit their scope. In parallel, the U.S. Army has attempted to answer some of the questions surrounding such model building, but for a different reason: Experiments that tracked infected animal populations during the 1950s and 1960s were designed to predict the effects of a BW attack.

And the development of increasingly sophisticated meteorological and vector growth models are areas of ongoing military interest. "Mathematical models have been developed to accurately predict cloud travel, area coverages, and dosages expected from a BW agent for a variety of environmental conditions," a 1985 Army report notes.

In 1983 Leonid A. Rvachev, a leading Soviet mathematical epidemiologist, sent several of his counterparts in the West a new series of algorithms he had developed. They constitute a model, Rvachev wrote, for accurate prediction of a worldwide influenza pandemic. The model combined many disease and population factors with data on climatic conditions and transportation routes into a formula so complex that only a handful of scientists anywhere could fully comprehend how to apply it.

According to his correspondence, Rvachev hoped that openness would diminish the prospect that the model could become the first comprehensive recipe for solving a seemingly insoluble problem of BW control: how to forecast the spread of a disease over areas as wide as continents, indeed, over the entire earth. The lack of confidence in such predictions has always made those with their fingers on the BW button queasy.

"The document itself is not worrisome," according to Ira Longini, an expert in epidemiological statistics at the University of Michigan. "What is worrisome is Rvachev's letter," which suggests that in the wrong hands the model could lead to biological warfare. "He wishes to help establish a surveillance system such that neither side will be able to exploit the methodology toward weapons use."

THE QUESTION OF PLAUSIBILITY

In considering the arms control implications of biotechnology we must draw a clear distinction between plausible *applications* and realistic enhancement of the ability to wage biological or chemical war. This much is clear: Many of the applications mentioned above are already accomplished or soon will be possible. Some innovations could in principle be executed by the use of classical techniques, but rDNA allows more efficient or elegant improvisation. Others depend fully on rDNA.

The new biotechnologies can most certainly create incredibly frightful and dangerous organisms. But the impact of genetic engineering on the BW equation is far less certain. The weapons agents may create vast ecological and public health hazards without translating into the myxomatosis ideal.

Biotechnology is moving so fast that textbooks are obsolete before they go to press. Not surprisingly, a broad range of views about what biotechnology means for chemical and biological warfare exists in scientific, arms control, and military circles. Tables 4 and 5 show generally accepted views of independent (nonmilitary) authorities on BW and genetic engineering, drawn from personal interviews, and technical and popular writings.

TABLE 4
RELATIVE TECHNICAL FEASIBILITY OF INNOVATIONS IN BIOTECHNOLOGY RELATED TO CBW

NOW BEING DONE OR WILL BE POSSIBLE IN IMMEDIATE FUTURE	REASONABLY CERTAIN IN FORESEEABLE FUTURE	PROBABLY IMPOSSIBLE IN FORESEEABLE FUTURE
Increased virulence	Supertoxic toxins and chemicals	Highly selective ethnic weapons
Antibiotic/drug resistance	Increased controllability	Effective epidemic modeling
More efficient toxin production	BW agents with greater hardiness	Aerosol vaccination
Microencapsulation	New arthropod vector systems	Extremely broad-spectrum vaccines
BW agents that can defeat vaccines/natural immunity	Biochemical weapons	
BW agents that can inhibit diagnosis/countermeasures		
Improved production/storage		
Improved neurotoxin/CW antidotes		
Vaccines against most known BW and toxin agents		
CBW ultrasensors		
Safer lab/production conditions		

TABLE 5

RELATIVE LIKELIHOOD OF VARIOUS BIOTECHNOLOGY INNOVATIONS FOR MEANINGFULLY AFFECTING THE UTILITY OF CBW OR COUNTERMEASURES

IMMEDIATE OR SHORT-TERM INFLUENCE VERY LIKELY	INFLUENCE WITHIN TEN YEARS REASONABLY LIKELY	INFLUENCE WITHIN FORESEEABLE FUTURE UNLIKELY OR HIGHLY UNLIKELY
Antibiotic/drug resistance	BW agents with greater hardiness	Increased virulence
More efficient toxin production	New arthropod vector systems	Highly selective ethnic weapons
Microencapsulation	Biochemical weapons	Effective epidemic modeling
BW agents that can defeat vaccines/natural immunity		Increased controllability
BW agents that can inhibit diagnosis/countermeasures		Aerosol vaccination
Improved BW production/storage		Improved neurotoxin/CW antidotes
Safer lab production/conditions		Vaccines against most known BW and toxin agents
		Extremely broad-spectrum vaccines
		CBW ultrasensors
		Supertoxic toxins and chemicals

The very technologies that make many of these offensive applications plausible also render them self-defeating. Because effective defensive measures are virtually impossible, no nation could engage in a biological attack with the confidence that its own people were well protected.

When the ability to "plasticize" the antigenic structure of BW agents was discovered, vaccines and antitoxins became futile gestures. Even broad-spectrum vaccines would have to cover a limitless spectrum—a prospect no scientist is prepared to endorse—to achieve blanket protection against a reasonably predictable proliferation of recombinant threat agents. The inevitable result of a vaccine strategy is the biological equivalent of economic inflation: a frenetic, provocative race that can know no end.

Aerosol vaccination would be similarly doomed to failure. And it adds a range of unique technical problems that would make it ineffective in any case. No matter how densely an aerosol was applied, variable environmental conditions such as air currents would mean that many in the targeted population would not receive adequate doses.

Military planners might postulate that covert protection of a large population could be achieved with as little as 70 to 80 percent immunity because any outbreak of disease could be easily controlled under such conditions. Even so, individuals are likely to have widely varying anaphylactic responses to a new vaccine. Some, perhaps many, would inevitably suffer violent reactions, including death in extreme cases. Widespread problems of this kind would be difficult to conceal. Once alerted, immunized populations would be outraged and enemy forces would adjust their CBW strategies accordingly.

Ultrasensors also fall prey to the contradictions of vaccines. For a limited number of chemical agents, sensor alarm systems may be feasible under sympathetic environmental conditions. Systems that detect a range of chemicals could successfully warn soldiers even if they were uncertain about what chemical they were trying to avoid. But BW is another question altogether. Even if biosensors could be designed to detect 100 different organisms, they would be of dubious merit under conditions in which genetic manipulations may have created 200, 500, or 1,000 recombinant threat agents.

If there could be developed a generic system, one that would sound an alarm if any of a wide class of biological organisms were present, numerous false alarms would make it self-defeating. With each warning, the cumbersome, combat-inhibiting protective suits would have to be donned. An enterprising enemy might even be tempted to spray an area with innocuous organisms that would set off sensors to inhibit temporarily the movement of an opposing force.

A further problem with ultrasensors is their realistic scope. "Where and how would you use such a system?" asks Richard Novick, director of the New York Public Health Research Institute. Many ultrasensors would be required to secure even a small area confidently, he adds. To protect an entire army would be an astronomically expensive logistical nightmare.

Some other potential scientific developments appear to be even deeper in the realm of unlikely speculation. They attempt to achieve goals that are impossible or impractical in view of the vast complexity of our environment. Disease-spread modeling, for example, depends on conditions that are constantly changing—climate prominent among them.

Likewise, attempts to control reliably the action of biological organisms as suggested by Zilinskas, while possible in the lab, are by most realistic appraisals doomed to practical failure. "The extraordinary diversity, complexity and variation present in ecosystems makes [*sic*] it impossible to predict the effect of the agent," according to Jonathan King, professor of molecular biology at MIT and an internationally known expert in biological warfare. "Biologists can control the laboratory situation, but the environment is in continual flux. In addition, biological agents mutate and recombine with related organisms in the environment. Everything we know about the spread of infectious agents tells us that many other variables besides the strain of pathogen and strain of host affect the outcomes."

According to Frank Hoppensteadt, dean of the College of Natural Sciences at Michigan State University and an expert in mathematical epidemiology, the Rvachev model might have some predictive ability and could even be significantly refined. Hoppensteadt says reasonable prediction doesn't require "absolute reduction of all variables." It is a matter of how well a model is corroborated by experience. But he cautions that any such model contains far too many unknowns to be used with confidence.

The danger in epidemic modeling, Novick adds, is that it might contribute to false confidence among military leaders contemplating a BW attack.

Ethnic weapons may be an intriguing concept to a sinister mind, but their value is doubtful. Suppose, by the targeting of a combination of obscure genetic predispositions, there could be developed a recombinant BW agent that would reliably kill 85 percent of black Africans but spare 85 percent of whites. "If such a weapon were used by one nation against another of distinct racial composition, the resulting epidemic could spread to targeted racial types in neutral or friendly nations," Michael B. Callaham and Kosta Tsipis, arms control experts at MIT, wrote in 1977.

And using ethnic weapons would make the deniability factor exceedingly difficult to carry off. No plague has ever been known to kill widely and rapidly but with precise selectivity. The risk of an aggressor's being caught in the act and incurring international wrath would be unusually high, making the massive research and development effort ethnic weapons would require a dubious investment indeed.

The development of more virulent organisms, while technically possible, is unlikely materially to affect incentives or strategies for using BW. As fearsome as the visions may be, unstoppable supergerms are hardly necessary, considering the incredible range and potency of pathogens provided by nature.

By the same token, supertoxic chemicals and toxins will undoubtedly be feasible soon, but their advantages are again questionable. "I can't think of any agents that would be a real improvement unless you can get through the gas mask, and I can't think of any way to do that," says Harvard CBW expert Matthew Meselson. "It would take massive amounts of [a CW agent] to saturate the charcoal [in a mask's filter]. If you're going to do that, you might as well crush the guy under boulders."

And despite the apparent assets of superchemicals—reducing the size of unwieldy arsenals and easing dissemination—the increased toxicity correspondingly increases the problems. Such weapons would be incredibly hazardous to produce, transport, and deploy. Nerve gas, which has created a worldwide furor, is mere perfume compared with some agents on the drawing board. They would undoubtedly have to be prepared as binaries to reduce handling dangers, adding technical problems that might tip the scale against perceived benefits.

In view of all these shortcomings, what is left for CBW innovation? Three important weaponization factors—increased drug resistance, the ability to defeat vaccines and other medical countermeasures, and improved dissemination through microencapsulation—are available to provide almost immediate stimulus to the BW machinery. And if its technical problems can be solved, biochemical weapons may represent a new and important category. The other incentives introduced by new biotechnologies are in the realm of more efficient and effective production methods.

Biotechnology offers new options for guerrilla wars, as suggested in this book's opening scenarios. Covert sabotage of crops and animals could become increasingly attractive to any nation—or guerrilla army—with access to molecular biologists.

But rDNA will not lead to the "ideal" BW or routinize biological warfare. That would require a higher level of protection and predictability than is likely ever to be possible. Effective weapons will always pose deadly risks for their maker. And no realistic genetic transformation will yield biological weapons that are suitable for theater operations. Nor can biotechnology improve on the essential utility of chemical weapons in theater operations—surprise strikes against technologically advanced nations or, in limited cases, protracted attacks against primitively protected troops.

THE IMPORTANCE OF PERCEPTION

Many high officials of the Defense Department, State Department, and CIA publicly disagree with this appraisal. They say the Soviet Union will exploit every option, no matter how unlikely, to achieve its aggressive goals.

The Federal Emergency Management Agency (FEMA) and the Department of Health and Human Services, rather than the military, are responsible for protecting civilians against CBW. FEMA does not consider it "a major strategic threat to the U.S. population at this time" and therefore does not even have a program to deal with it. Neither agency is conducting research into countermeasures against BW agents, nor, say their spokespersons, has either agency ever asked the DOD for assistance. Regardless of the validity of the Soviet BW threat, Pentagon planners take everything seriously. In arms control this far outweighs biotechnology's "objective" impact.

To appreciate fully the significance of the Pentagon's fascination with biology, it is useful to consider the visionary speculation of Colonel W. D. Tigertt, former commander of the army's medical unit at Fort Detrick. "Those who would increase the potency of biological weapons must search for improved methods of mass production of organisms, factors which will enhance the virulence, ways to prolong the storage life of living agents, ways to improve aerosol stability, and methods of producing variant organisms by recombination or other means," he said in a 1963 article in *Military Medicine.* He also predicted mass aerosol immunization, ethnic weapons, and hormone weapons.

In short, Pentagon preoccupation with the very manipulations made possible by new biotechnologies is the product of decades of study and aspiration. From these roots has grown the DOD's apparently unshakable conviction that the key to biological warfare is at hand. Emboldened by certainty, this view has become increasingly rigid and hyperbolic.

Former Assistant Secretary Douglas Feith's belief that with biotechnology "new agents . . . tailored to military specifications . . . can be produced in hours" may have overstated even the DOD's case. But it is consistent with the Reagan administration's general outlook. "The area of biotechnology has the potential to have a major impact on national security," concluded the Defense Science Board on Chemical Warfare and Biological Defense in 1985.

In the same year the president's Chemical Warfare Review Commission said the DOD "does not have an adequate grasp of the biological warfare threat and has not been giving it sufficient attention. Both intelligence and research . . . are strikingly deficient. The Department should be devoting much more resources and talent to addressing the chemical and biological threats of the future as well as those of the present."

"Perhaps the most significant event in the history of biological weapons development has been the advent of biotechnology," the army reported to a House committee in 1986.

The message is that biological weapons are viewed as important, "surgical-strike," tactical weapons—instead of just second-rate, indiscriminate weapons of mass destruction—for the first time in decades. Regardless of stated concerns over the Soviet threat, these officials know—or certainly should know—that the use of chemical or biological weapons by major powers against one another would be fruitless at best and could provoke a nuclear response. So if biotechnology is sent to war, it will be by the superpowers against third world targets.

"U.S. armed services have no actual experience of using poison gas in war for 64 years," SIPRI noted in 1982. "For the United States, then, notions of what the latest types of poison-gas weapons can do, how many of them are needed to do it, and how they relate to other weapons, are now entirely theoretical: inferences from history, from field experiments and from computer simulations that can reflect the likely realities of future combat only dimly. These are the notions, however, from which the formal military requirements for poison gas are stated."

Based so heavily on hypothesis, expert advice on America's chemical arsenal fluctuates between vast increases and dramatic cuts. To "play it safe," the Reagan administration has always fought for increases.

The DOD's inflated view of the importance of biotechnology has a great deal to do with this power of abstraction. CW planners have at least the World War I experience to draw upon. But no nation has the benefit of knowledge gained from extensive large-scale BW attacks. Even biological field testing has been eliminated in recent years.

Genetic engineering, having emerged only in the 1970s, is being evaluated for its weapons potential in an experiential vacuum. Imaginations have run wild. In the past, when improvements in the technical performance of weapons systems were demonstrated, those changes were almost always made. In this way biotechnology exerts tremendous pressure on arms control.

THE OFFENSE-DEFENSE TAUTOLOGY

The DOD has often stated that its biotechnology research, like the rest of its BW program, is completely and properly defensive in nature. In light of the above concerns and a generally unstable arms control climate, can that claim be taken at face value? The answer lies largely in the relationship between offensive and defensive efforts in biological warfare.

To a degree greater than other weapons systems, nearly all aspects of offensive and defensive research and development pertaining to BW are identical. Even the military makes no attempt to dispute this fact.

"It all comes down to intent. . . . With technology plus intent, you can do great good or great harm," William Beisel, former director of BW research at Fort Detrick, said in 1984. "We're the good guys; we're wearing the white hats," he added. "We're physicians using [rDNA] to create new things in medicine. It's the Russians who may be using [rDNA] for evil."

So reassuring a statement deserves a second look, coming as it does from one of the more Strangelovian characters the army has produced. Beisel presided over the testing of offensive BW agents on the volunteers of Operation Whitecoat before it was discontinued in the early 1970s.

"What we were to learn this morning was that we would be injected with endotoxin [a bacterial toxin]," read the letter of one Whitecoat participant. "This time both a nurse and Lt. Col. Beisel of the Army Medical Corps were present at the injection. He injected the needle deep into my vein and told me that shortly I should have some reaction. . . . Within an hour the top of my head felt like all the gremlins in Hades were inside trying to emerge by hitting the underside of my skull with sledge hammers."

As Beisel's statement demonstrates, defensive efforts predicated on mistrust are sure to lead to mistrust. It would hardly be surprising if given the mistrust they could logically be expected to feel for the Dr. Beisels of the United States, the Soviets were boosting their BW research program.

The problems with defensive research, however, go beyond doubts about good faith. The concept of "defensive" research is itself a fundamental misrepresentation. Table 6 summarizes the basic features of BW research.

TABLE 6

FEATURES OF A BW RESEARCH PROGRAM

Microorganisms: **Basic and applied research on a wide range of candidate agents.**
Test facilities: **Highly contained, access to animal models.**
Delivery vehicles: **Ability to test some combination of arthropods, bombs, missiles, aerosols, or water contamination.**
Countermeasures: **Development of vaccines, antibiotics, and other prophylactic or therapeutic measures; protective clothing and other physical barriers; well-trained personnel for detection and protection; decontamination procedures.**
Detection: **Development of lab and field methods for detection/confirmation as well as alarm systems.**

Each feature is essential for both developing and protecting against BW. Viral vaccine research and development, a major thrust of the DOD biotechnology program, demonstrates this point. Before any existing viral pathogen or newly created mutant could be manufactured as a weapon, a vaccine must be developed to protect troops that would use the weapon as well as researchers and technicians who would handle it in the lab and mass-produce it. In the case of a "defensive" vaccine program, a nation would naturally try to target viral agents suspected to have been developed as weapons by an enemy.

"The development of the [defensive] vaccine will in general involve the isolation, identification, modification, and growth of the potential BW agent," according to an article by King and his MIT colleague Harlee Strauss. "Studies of the vaccine's efficacy would have to be carried out, requiring exposing test animals and perhaps humans to the agent. These steps—the generation of a potential BW agent, development of a vaccine against it, testing of the efficacy of the vaccine—are all components that would be associated with an offensive BW program. It is not that the programs 'appear' similar; it is that they share similar components."

If a research team looked at a virus for such factors as its biochemical mechanism of action, the routes through which it is infective, whether it can be made resistant to drugs, and its stability in storage, this work might be considered defensive. How can an agent be defended against if its essential characteristics are unknown? Yet these are also major factors for any offensive inquiry.

If a vaccine were successfully developed, a defensive program might then move to mass-produce and stockpile it, as the United States has done in a number of cases. Clearly any nation planning a BW attack would do likewise.

But this is where the logic of defense doubles back on itself: The value of vaccines is negligible. Even before rDNA technology increased the number of potential threat agents exponentially, it was impossible—for medical, logistical, and financial reasons—to vaccinate all soldiers and civilians against all potential BW agents.

And once used, vaccines invariably have an incubation period before they confer immunity. Many vaccines cannot be easily stockpiled in large quantities or administered on short notice. For vaccines to be used effectively, the defenders would have to know with absolute certainty which agents would be used weeks or even months in advance of the attack—obviously impossible predictions.

At the same time BW vaccine research is inherently suspicious. Vaccine manufacturing may easily camouflage BW production. Although "the typical vaccine plant is inadequate for the production of a fully military capability, it would be adequate for the production of the quantities required for a sabotage attack or the kind of low-grade capability that might be sought by a country in the face of military defeat or frustration," a SIPRI study noted in 1972. "Some common forms of vaccine production are very close technically to the production of CBW agents and so offer easy opportunities for conversion." These observations are even more meaningful today, in light of the advances in bioprocess technologies.

It's worth noting that the Japanese BW program during World War II was operated under the guise of "water purification" and "vaccine production." Although the American occupation force helped keep the program secret, for more than thirty years these labels lent credibility to Japanese denials of their heinous attacks on the Chinese and experiments on human guinea pigs.

"A defensive [CBW] program not supported by an offensive program can well be worthless," General William Creasy, head of the U.S.

Army Chemical Corps, told a congressional committee in 1958. "You cannot know how to defend against something unless you can visualize various methods which can be used against you, so you can be living in a fool's paradise if you do not have a vigorous munitions and dissemination-type program."

Creasy's basic position—that the study of prototype agents and munitions is essential to a legitimate CBW defense—has been adopted by his successors. The words *prototype* and *new biological weapon* tend to blanch the average citizen and thus have been avoided by DOD spokespersons in recent years. Yet in a multitude of government reports, interviews, and congressional hearings, Creasy's position has consistently been quietly affirmed.

"The characteristics of classical agents are fairly well known but we do not have similar critical information on the emerging threat," the army reported to a House of Representatives committee in 1986. "We especially need more information about protection against novel agents."

"There's no other way to build defenses unless you have the agents," said Colonel David Huxsoll, commander of the Army Medical Research Institute of Infectious Diseases (AMRIID), in agreement.

The necessity of prototype development, inextricable from the military's view of effective defense, certainly represents the best reason to fear that offensive goals already complement and will ultimately supplant the high-minded search for medical countermeasures.

Stocks of BW agents held by defensive programs are also cause for concern. To understand the properties of an agent in order to defend against it, significant quantities must be tested, particularly when its potential for aerosol dissemination is verified. "It may be possible to discern the difference between offensive and defensive intent by the size of the manufacturing and/or dispersal facilities," King and Strauss have pointed out, "but usually there will be a large gray area where claims of either intent will be equally convincing."

In the case of relatively small-scale sabotage attacks, this becomes a fundamental distinction; for example, efficient aerosol applications of a single kilogram of Q fever culture or as little as seven grams of purified botulinal toxins could have a direct military effect over a square kilometer. Similarly small quantities of other organisms could be used to poison large water or food supplies.

Evaluating the intent of BW research is further complicated by its potential link to important public health and biomedical research questions. The study of disease is hardly the exclusive province of the military. By the same token, academic and commercial findings are available for military application. Agricultural research, for example, now involves attempts to stabilize recombinant organisms to withstand adverse environmental conditions and aerosol dispersal.

And offensive-defensive parallels extend far beyond research into threat agents per se. Defenders need sophisticated detection and decontamination methods to determine properly when an attack has taken place and what organism has been used in order to mobilize appropriate countermeasures. The same capabilities are vital to an aggressor "in order to determine the extent of spread of the biological agent and when an area is safe for the occupation of a non-protected population," King and Strauss have stated. "Methods of decontamination are important to rid an area of the harmful organism so it can be reoccupied" as well as to clean up spills of hazardous organisms during research and development.

Training for biological warfare is similarly ambiguous. In both cases, troops must be instructed in decontamination and the effective use of protective garments and masks. To mount a biological invasion, only a small number of additional troops would require training in munitions handling.

Likewise, well-coordinated command and control systems would be needed for nations on either end of a biological attack—to predict its effects and to determine and carry out offensive initiatives or protective actions and countermeasures. Defensively or offensively officers expert in biological warfare would take charge.

THE FALLACY OF MEDICAL DEFENSE

An important benchmark for evaluating the worth and motivation for medical defense programs is the fact that societies are rarely able to control the spread of even naturally occurring organisms. Tens of thousands of people stricken with the flu each winter, regardless of public health efforts, can attest to this.

Some programs are highly unlikely on the basis of cost alone. It's not inconceivable, for example, that a country might try to combine medical countermeasures with protective shelters against CBW attacks, as has been attempted by some nations for nuclear fallout. In the early 1970s, according to a UN estimate, effective shelters for a developed nation of 100 to 200 million people would cost $15 to $20 billion. Today the figure could easily top $100 billion.

Medical defense, the linchpin of any BW defense program—as the vaccine strategy shows—is not merely expensive and futile but ultimately also self-defeating and destabilizing. And as fruitless as medical defense may be for human beings, there is no realistic defense of any kind for crops or animal targets.

Perhaps more important, as the DOD's Feith has pointed out, "The BW field favors offense over defense. It is a technologically simple matter to produce new agents but a problem to develop antidotes." As Table 5, showing the relative effects of technical innovations on CBW, indicates, biotechnology will only *extend* and *accelerate* this disparity. Not a single plausible application is likely to increase meaningfully the prospects for effective CBW defense.

DOD views on BW defense tend to be contradictory and muddled. Its affirmation of the relative advantage of offensive research over defensive notwithstanding, the Pentagon also states a firm belief in the legitimacy of medical defense, as the massive effort described in the next chapter shows. This categorically includes the need for testing actual BW agents and stockpiling vaccines in the millions of doses, according to AMRIID commander Huxsoll, who, more than any other individual, determines the course of this research and development program.

In the era of rDNA General Creasy's model would dictate the development of prototype recombinant BW agents to test and challenge countermeasures. But when pressed, officials distance themselves from such an appraisal. Following are excerpts from an extensive interview conducted by one of us (Piller) with Huxsoll in 1986:

PILLER: "Regarding physical and medical countermeasures, presumably in many cases the agents themselves, at least on a laboratory level, must be studied."

HUXSOLL: "No question about that."

PILLER: "Does this imply that at some point genetically altered organisms that are hazardous, and potential BW threats, must be created in some kind of prototype form for defensive research to explore what the countermeasures might be?"

HUXSOLL: "Generally that's not the case. You don't necessarily have to go around altering organisms to build a defensive capability. . . . That would be an endless effort, and I think to no avail . . . [to] develop a defensive measure for every conceivable alteration, I think, would be ludicrous. . . . We look at antiviral drugs, immunomodulators, immunopotentiators, those kinds of things. We're using a generic approach to protection. A generic approach means making a common drug, or a vaccine, that might be effective against a broad group of agents. . . . By looking at things like antiviral drugs that act in such a way on the virus or cell that it would make overcoming the efficacy of that drug extremely difficult. In other words, it may be virtually impossible to alter that [viral BW agent] to the extent that it's no longer susceptible to the antiviral drug."

This generic or broad-spectrum drug and vaccine approach, as noted earlier, is not complete nonsense. But virtually no qualified independent scientist believes it will lead to an acceptable level of medical protection against even *known* BW agents and toxins, let alone thousands of recombinant threat agents designed specifically to defeat drugs and vaccines.

And even within Huxsoll's strategy, DOD policy is vague or nonexistent. Any effective use of vaccines must take place weeks or months ahead of exposure to the BW agent. But the following comment demonstrates that there are no apparent guidelines that define when the stockpiles of millions of doses of defensive vaccines might be used:

PILLER: "Is the goal to vaccinate U.S. troops on a mass scale, along the lines of these generic or broad-spectrum vaccines?"

HUXSOLL: "OK, some of the definitive policy is not [yet developed] on use of vaccines that may be important as a protection against potential BW agents. . . . How do you know what the trigger is? A change in intelligence information so that it might become information that is quite definitive [could be the trigger]. You might have a situation where a given unit is exposed, and the reaction might be, 'Well, they got us once, but we're going to protect ourselves, and they're not going to get us again.' But if the question is that if we develop fifteen vaccines, are we going to vaccinate every soldier against fifteen different agents, on top of what they're already getting, that answer is probably no."

Absent from Huxsoll's explanation is any logical analysis of central problems in BW defense. He acknowledges that the vaccine program is proceeding apace but is rudderless. Policies to dictate when or if its products should be used have yet to be established. More significant, perhaps, is his idea that vaccination might prevent the enemy from "getting us again." It seems incredible that the commander of the largest U.S. BW research institute would not have surmised that the enemy might shift to a different BW agent when dozens, hundreds theoretically, are available.

A further example of apparent confusion on the part of U.S. BW planners is the continued pursuit of aerosol vaccination strategies. For very sensible reasons related to efficacy, Huxsoll himself rejects the method for mass vaccination of soldiers or anyone else. At the same time his institute continues to study the technique intensively. Even as he discounted the method, Huxsoll said: "In some places, including the Soviet Union, they do a lot of that; at least I'm told they do."

Huxsoll's explanations are so awkward and seemingly contrived that his program's goals must be questioned. Is the Pentagon's "defensive research" capitalizing on the benefit of the doubt?

THE DANGERS OF DRIFT IN DEFENSIVE PROGRAMS

In view of the military's vague, often contradictory separation of offense and defense, the political and bureaucratic milieu of BW research and development plays a pivotal role in the direction and propriety of the work. We address the integration of independent scientific expertise into military biology in Chapter 9, but it is also relevant here.

"The attitude of scientists toward CBW certainly depends in most cases on the context in which their judgement is sought," a SIPRI study observes. If they perceive a legitimate need for defensive research, many will come forward. "Thereafter, by small and perhaps imperceptible stages, individual scientists or research laboratories may drift over from defensive activities into offensive ones, even though they were not originally recruited for that purpose (e.g., from prophylactic medicine . . . into tests with nonpathogenic simulants and animal pathophysiology with potential weapons . . . to weapons development to delineate the possibilities open to potential enemies)."

This progression is particularly likely when scientists are wooed into large military research institutes where a group identity is easily forged and some significant preventive medicine achievements normally are spun off from the work. As these health-related by-products are published and promoted within the civilian academic community, they contribute to a breakdown of the separations between BW researchers and public health scientists.

Fort Detrick in the late 1940s was the archetype. The BW labs there acquired a status within the Society of American Bacteriologists (SAB) akin to that of a leading university that trained young scientists and made significant contributions to basic research.

"To a large extent, those who participated directly in [Fort Detrick] activities did so because they felt that their country was in a crisis that was expected to pass in a limited time," the SIPRI analysis continues. "When this did not happen, qualms of conscience began to set in: could there be such a thing as an infinitely sustained crisis, and was continued biological-weapon R&D morally permissible?"

Ultimately this contradiction, and concern over secrecy and professional ethics, disrupted the cozy relationship between the SAB and Fort Detrick. In the late 1960s, as the cold war ended and the CBW program fell out of popular and political favor, the SAB moved to distance itself from military research.

As this example shows, the relationship between the military and the civilian biological research sector on which it depends for scientific expertise is not fixed. When the relationship is close, as was the case in the late 1940s and early 1950s, its mantle of respectability inevitably promotes a drift of civilian and defensive military interests into the offensive realm.

Three current conditions (explained fully in later chapters) are replicating those days of substantial military-civilian cooperation and offensive drift. First, the DOD is funding civilian biological research at unprecedented levels. Military support is gaining a respectability in the life sciences it has rarely known. Secondly, with the computerization of scientific data and communications systems, civilian-military cross-fertilization occurs with increasing ease. Finally, the Soviet specter has been evoked more stridently in this decade than at any time since the cold war. The resurrected BW threat stalks the halls of science once more. Civilian scientists are again being admonished that their country is in a crisis and needs their help.

BW AND NATIONAL SECURITY

"Since the first atomic bombs were dropped on Hiroshima and Nagasaki in 1945, military theorists have recognized that there is a nuclear 'firebreak'—a barrier that will prevent even the most intensive forms of conventional combat from escalating into nuclear war," according to military analyst Michael Klare. "Existence of the firebreak attests to the fundamental, qualitative difference between the most destructive forms of conventional warfare and the catastrophic potential of nuclear conflict. However violent a conflict on the conventional side of the barrier, it cannot annihilate the planet."

That essential psychological deterrent to total destruction, Klare believes, is now in grave danger. "Recently . . . the development of new types of both conventional and nuclear munitions has begun to erode the firebreak. Modern conventional weapons employ explosive technologies that begin to approach the damage potential of the smallest nuclear weapons. And nuclear munitions now under development have the controlled destructive effects associated with conventional arms."

As we've seen, some third world countries are apparently seeking CBW as a strategic alternative to nuclear weapons, which are beyond their economic-technological reach. Nations prone to this posture are slow to grasp a hard reality: Because of virtually insurmountable problems of large-scale delivery, control, and reliability, CBW could never credibly claim to fill the role of H-bombs.

But increased worldwide interest in CBW can still have ominous consequences. "So-called limited wars are fought with conventional weapons which individually affect only a small area," Meselson told a Senate committee in 1971, in support of U.S. ratification of the 1925 Geneva Protocol. "Such weapons can be decisive only when very great quantities are available. The wealth of the United States allows us to expend enormous quantities of conventional munitions in tactical combat. Very few countries even approach this capability. However, the proliferation of lethal chemical weapons would greatly enhance the destructive capability of smaller and less wealthy nations."

Using CBW would by no means ensure wartime victory for developing nations—particularly if proliferation became widespread. But their wars would be far more deadly. The nuclear firebreak would be perilously narrowed.

"In the context of both tactical and strategic war, it is very much in the United States' interest to preserve and strengthen the restraints that prevent chemical warfare and the proliferation of chemical weapons," Meselson said at the Senate hearings. His comments apply equally to BW. In a 1986 interview Meselson added:

> Failure to understand the history and logic of the U.S. renunciation of biological and toxin weapons can lead to a serious error. It's sometimes said that if new developments in molecular biology make biological weapons more effective and easier to produce, they might become more attractive to the U.S. Exactly the opposite is true. [If such improvements were possible, they] would make even more compelling the arguments for the existing U.S. policy of nonproliferation [of BW and toxin technology], of which the credibility of our own categorical renunciation is the most essential element. Those who do not understand this play a dangerous game, which could cause great harm to the United States.

As we have shown, for reasons of science and simple logic, a BW defense program both is doomed to failure and contains inescapable, highly provocative offensive applications. Yet the Pentagon takes the implications of rDNA so seriously that it continues to conduct a massive BW defense apparatus. In so doing, is the United States compromising "the credibility of our own categorical renunciation" of BW? Is the U.S. military enhancing the attractiveness of BW to all other nations?

If so, it is undercutting U.S. national security. Ultimately it may make little difference whether the United States calls its BW program "defensive" or "offensive." In either case it is dangerously misguided.

6

APPRAISING THE U.S. MILITARY BIOTECHNOLOGY PROGRAM

THE PENTAGON'S FRAMEWORK

The Defense Department's program in biological warfare defense has never faced the kind of detailed public scrutiny that many government programs are often subjected to. However, a great deal is known about its basic structure, stated goals, and methods. Most research and development fall into one of five areas: medical defense, protective gear, decontamination technology, ultrasensors and other devices to detect or measure BW agents, and intelligence gathering.

The military conducts basic research in the life sciences—such as molecular biology and environmental biology—to maintain a technology base for certain nonmedical aspects of BW defense, including ultrasensors and decontamination.

Basic and applied medical researchers investigate well-known and newly discovered organisms or toxins to uncover their properties and dangers. This involves analysis of everything from the genetic makeup and virulence of threat agents to development of methods for their isolation, large-scale reproduction, and dissemination. Ultimately they aim to create and test vaccines, toxoids, or drugs to counteract or neutralize agents that are seen as threats to U.S. forces. These products are often manufactured on a mass scale. Vaccines for a range of diseases have been produced in the millions of doses.

The army is the lead agency for all BW work. Its Medical Research Institute of Infectious Diseases, at Fort Detrick, is the heart of BW medical defense, while its Test and Evaluation Command runs the physical protection program. The navy also conducts substantial BW research. All the military agencies involved in this research contract out a large portion of their work to universities and corporations all over the United States and to a limited extent in allied nations.

BW defense is a prideful program. In addition to playing a vital role in protecting the nation, its leaders say, it contributes broadly to society. In 1971 an epidemic of Venezuelan equine encephalomyelitis (VEE) ravaged northern Mexico and Texas, killing hundreds of horses and infecting scores of people. The vaccine that quelled the outbreak was furnished by the BW program. "It's the only effective vaccine against VEE in the Western Hemisphere," an army spokesman said at the time. "It can be considered a beneficial result of biological warfare research."

And DOD representatives point out that BW researchers have made important contributions to the understanding of fundamental questions of disease transmission. The study of antiplant and antianimal weapons has occasionally led to agricultural advances. More effective detection of BW agents has aided in the control of environmental pollutants.

These civilian spin-offs—though incidental to military goals and accomplished far less economically than projects designed to serve civilian goals directly—add a layer of complexity to the justification for biological warfare research. But the fundamental rationale for this massive effort is simple: the Soviet threat.

The U.S. military justifies its actions by the firm conviction that the Soviets are ruthless criminals. The BW program has grown robust in the lengthening shadow of this fear. Any credible analysis of U.S. goals and intent must recognize its far-reaching influence. The overall impact of U.S. perceptions of the Soviets is addressed in Chapter 10. But the comments of one key BW scientist are a fitting preface for a description of the U.S. biotechnology program.

"This lab and the [AMRIID] are charged with a very frightening task," said David Kingsbury in 1984, when he was director of the Naval Biosciences Laboratory in Oakland, California. "We're the two Defense Department labs that are tasked with identifying biological warfare agents. That's one of the most awesome tasks I can think of, coming up with a definitive statement that we've been attacked with a biological weapon, knowing that that statement is probably equivalent to pushing the [nuclear] button. Reagan could always call the Kremlin and ask, 'What the hell did you do that for?' My guess is he wouldn't. He'd tape that message to the front end of a Minuteman missile."

Kingsbury went on in 1985 to take a new job at the National Science Foundation. There he serves as one of the leading regulatory policymakers on rDNA technology.

SIFTING THE FACTS

The analysis that follows is the product of years of investigative research. To appraise the role of biotechnology in U.S. biological warfare efforts, we reviewed only open information. We did not have access to classified materials.

Our review is based in part on the DOD's "Annual Reports on Chemical Warfare-Biological Research Program Obligations," known as "obligation reports." In 1976, in the wake of revelations of unauthorized activity in CBW, Congress mandated that all unclassified information from the Pentagon's CBW work be summarized in these reports and made available to the public. They contain a broad, though sketchy, overview of trends and priorities in the field.

We also used a wide range of technical reports, analyses, and academic publications generated by the military as well as promotional information about the BW research program. These were complemented with press reports and extensive interviews of DOD officials and independent experts.

Over a period of four years one of us filed more than a dozen requests under the Freedom of Information Act (FOIA) for documents on military genetic engineering. The most fruitful result was the release of more than 300 "DOD Research and Technology Work Unit Summaries," one-page forms prepared for each DOD project in all fields. They contain brief descriptions of the military objective and technological approach of the researchers as well as a progress report or final statement.

Work unit summaries are uniquely valuable. They show the direct connection between research and military mission, are created for each project, are succinct enough to review by the hundreds, and are sometimes accessible to dogged investigators.

This analysis is thorough and rigorous—but unfortunately not comprehensive. After describing the DOD's biotechnology program, we explain how the military penchant for secrecy has seeped into areas mandated to be open to public view. The result has been to veil, if not to black out, any detailed independent analysis of the BW program—to the peril of U.S. and international security.

We analyzed 329 separate projects funded from 1980 to 1986, for which we obtained work unit summaries or other clearly identifiable, specific documentation. All were designated by the DOD as directly related to biotechnology. Although this represents the most detailed, comprehensive independent overview of DOD work in biotechnology, it is nevertheless a *nominal* analysis. Several key avenues of research noted in alternate DOD sources are barely mentioned in the work unit summaries that we obtained. Therefore, the 329 projects are unlikely to portray a representative cross section of the CBW-biotechnology effort as a whole.

We did not attempt to infer priorities for specific projects or general topics from the total funds allocated for each—ranging from a few thousand to several million dollars—although that information is included in our data base. This is because cost is not always a good indicator of relative importance in biotechnology and because the military refused to release the full complement of supporting documentation.

Although the research and development priorities cannot be precisely quantified, when they are supplemented with more cryptic information from other DOD sources, conclusions can be drawn about the apparent implications of the total body of work. Table 7 presents an overview of the 329 core projects.

TABLE 7
INSTITUTIONS, GOALS, AND METHODS OF 329 DOD BIOTECHNOLOGY PROJECTS

Institution Conducting Research/Development Project

Total projects	329	100%
DOD in-house lab	143	43%
Total contractor	186	57%

Academic	166	51%
Corporate	16	5%
Governmental	4	1%
Stated Goals of Research/Development Project		
Vaccine Development	61	19%
Toxin, antigen isolation/characterization	102	31%
Development/use of antibodies as therapeutics	36	11%
Diagnostics/ultrasensors	66	20%
Manipulation of antigens	12	4%
Aerosol vaccination	2	1%
"Basic" studies related to cell growth, immunity, neurobiology	61	19%
Other "basic" studies	46	14%
Antiviral drug development/characterization	9	3%
CW antidote development	19	6%
Broad-spectrum therapies	12	4%
Fellowship/equipment/conference funding	43	12%
Clinical research	23	7%
Stated Research Methods		
Recombinant DNA	107	33%
Monoclonal antibodies	126	38%
Microencapsulation	2	1%
Manipulation of antigens	3	1%
Computer modeling	6	2%
Classical biochemistry/cell cultures/genetics	125	38%
Bioprocess technology	4	1%

Note: Some percentages do not total 100 because of rounding and multiple applications for a single study; some study counts exceed 329 because of projects that have multiple applications.

We also appraised the apparent scientific merit of each project along with its potential for offensive use. The process was subjective, especially in view of the limited amount of available information, but it employed well-established standards of quality and goals for biomedical research, based on years of experience writing and reviewing grant proposals for the National Institutes of Health, the National Science Foundation, and other agencies.

Somewhat to our surprise, the projects fell into three simple, overlapping primary categories:

- *Offensive:* research that has obvious *potential* for offensive applications. (None of these studies was labeled "offensive" by the DOD, of course. That would be an acknowledgment of violations of U.S. policy and international law.) This category constituted about one-fifth of contractor projects and fully one-third of in-house military projects.
- *Standard:* research similar to projects that might be submitted to the NIH. Many studies in this category, however, appeared inferior to those typically funded by the NIH. This category constituted about seven-tenths of contractor studies and less than half of in-house military work.
- *Poor:* research that is poorly conceived or designed, uses confused logic, or addresses problems of questionable significance. This category constituted about one-fifth of contractor work and one-fourth of in-house military projects.

We then created a list of offensive applications of biotechnology, based on the factors detailed in Chapter 5. Many studies in the "standard" and "poor" categories had clear potential for offensive application, but we discuss here only the eighty-six studies that seemed most explicitly "offensive" in nature. This eliminated studies whose offensive applications are obscure or could be considered extraneous or an abstraction of the overall research goals. It concentrated and refined the sample. These were then categorized by specific applications in Table 8.

On the basis of this initial review, offensive projects appear to emphasize the creation of novel BW agents, agents that defeat vaccines and inhibit diagnosis, new methods of dissemination using biological organisms as the vehicles (vectors), and the increased production of toxins, especially those with enhanced effects.

At face value, these are very intelligent priorities for an offensive BW program. Regardless of the Pentagon's motives or perspectives on the use of biotechnology, these projects would likely have the greatest chance of near-term success. Defeating vaccines, for example—a frightening BW enhancement that is quite feasible with today's biotechnologies—appears to have been an area of intense scrutiny. In contrast, more fantastic prospects that pose forbidding technical and logistical challenges, such as ethnic weapons, apparently received little attention.

To extend the analysis, we then looked at the four *stated goals* appearing most frequently among the eighty-six studies—vaccine development, toxin/antigen isolation and characterization, development and use of antibodies as therapeutics, and diagnostics and ultrasensors—and

TABLE 8

POTENTIAL OFFENSIVE APPLICATIONS OF EIGHTY-SIX DOD BIOTECHNOLOGY PROJECTS

POTENTIAL OFFENSIVE APPLICATION	NUMBER	PERCENTAGE
BW agents that defeat vaccines	23	27
BW agents that inhibit diagnosis	14	16
Supertoxins*	17	20
Aerosol delivery of BW agents	5	6
Biological vectors for BW agents	19	22
Novel BW agents	51	59
Drug-resistant BW agents	3	3
Highly specific ethnic weapons	0	—
Biochemical (hormone) weapons	1	1
Increased toxin production capability	15	17

*Genetic alternations that could increase the toxicity of an organism's toxin products.
Note: Percentage total exceeds 100 and study count exceeds 86 because of multiapplication projects.

compared them with the list of potential offensive applications. The results of these comparisons were fully consistent with the initial observations.

More significant, however, this breakdown gives an important clue to how an offensive program may be "coded"—what each defensive-sounding label on the DOD work unit summaries may represent. Following are the *offensive* applications that might lurk beneath the four major defensive *stated* goals:

Vaccine development

Novel BW agents
Defeat vaccines
Increased toxin production
Supertoxins

Toxin, antigen isolation/characterization

Novel BW agents
Defeat vaccines
Increased toxin production
Supertoxins
Biological vector delivery

Diagnostics/ultrasensors

Biological vector delivery
Novel BW agents
Defeat vaccines

Development/use of antibodies as therapeutics

Novel BW agents
Defeat vaccines
Inhibit diagnosis

Although these designations are based on limited data, they show logical applications of the DOD's studies to an offensive program. Therefore, a review body—be it an independent panel of concerned scientists, an official agency of the government, or an international arms control team—could use these relationships as a point of departure for probing the motives of the overall DOD program in biotechnology as it develops over the next few years.

When we use these "codes," some relationships between otherwise dissimilar, geographically and institutionally separated projects seem to emerge. Biotech Research Laboratories, a small company in Maryland, won a DOD contract to produce monoclonal antibodies to surface antigens of the bacterium *Bacillus anthracis* (the organism that causes anthrax) to "support ongoing studies of infectious diseases." Meanwhile, an investigator at Louisville University obtained contract funds to test the reactivity of lectins, naturally occurring compounds that can target and bind tightly to specific molecules in bacterial cell surfaces, "to determine if different lectins have different affinities for various microbes." This work showed that *Bacillus anthracis* was selectively bound by certain lectins. While there is no stated relationship between these projects, one wonders whether the bound lectin would block access of the monoclonal antibody to the *Bacillus anthracis* cells. Such a discovery could easily find offensive application since it would allow potential pathogens to escape detection.

As a second example, Stanley Falkow, a noted Stanford University researcher, obtained an army contract to clone a gene from a pathogenic bacterium into the innocuous bacterium *E. coli*. The novel *E. coli* strain gained the ability to attack human cells in the same manner seen for the pathogen. Meanwhile, a group at the Walter Reed Army Institute of Research developed methods to modify genetically bacteria such as *E. coli* "to produce any desired antigenic structure and level of pathogenicity." Obviously, joining these projects might yield a new generation of invasive and deadly BW agents.

Thus, even research that may seem harmless or even scientifically meritorious in isolation must be viewed warily, considering the possibility that Pentagon planners are combining the fruits of labor from the broad array of work the army supports.

Still, some areas, such as aerosol delivery of BW agents and microencapsulation—feasible and logical needs for any offensive effort—were strangely absent from our sample. Does this mean they are being ignored by the military?

Hardly. The DOD obligation reports fill in many of the apparent gaps. Here are some areas of research and development—all essential to a well-coordinated offensive program—that are indicated as significant priorities in these reports but were not well represented in the 329 core projects:

- Microencapsulation
- Aerosolization of BW agents and toxins to evaluate offensive potential and protective gear
- Mathematical models for aerosol delivery
- Aerosol vaccination methods
- Aerosol delivery of drugs and nerve agent antidotes
- Industrial-scale bioprocess facilities
- Vaccine stockpiling in the millions of doses
- Broad-spectrum and synergistic antiviral drug development
- Insect vectors

And the DOD response to a 1986 lawsuit filed by the Washington, D.C.-based Foundation on Economic Trends, revealed additional studies on target vulnerability in Europe, Asia, and the United States and on "Biological Agent Delivery by ICBM (intercontinental ballistic missile)."

THE CONSISTENCY FACTOR

This range of work clearly indicates pursuit of the most obvious potential offensive research and development applications of the new biotechnologies. Table 9 shows how the DOD's current program might fit into a well-coordinated, deliberate strategy to develop an offensive capability.

TABLE 9
U.S. BW PROGRAM DEVELOPMENT—OFFENSIVE IMPLICATIONS*

STAGES IN THE DEVELOPMENT OF AN OFFENSIVE CAPABILITY†	VERIFIED EVIDENCE	SOURCE
1. Policy review of pros and cons of offensive BW program	Yes	b
2. Decision to proceed on offensive program	No	—

U.S. BW PROGRAM DEVELOPMENT—OFFENSIVE IMPLICATIONS (*cont.*)

3. Preparation of detailed budgetary estimates for R&D	Yes**	b
4. Voting of R&D budgets	Yes**	b
5. Recruitment of R&D personnel	Yes**	b
6. Organization of in-house research facilities, funding of university/corporate researchers to build flexible technology infrastructure	Yes	b
7. Selection of BW agent research projects	Yes	a,b
8. Selection of BW agent development projects	Yes**	a,b
9. Organization of development facilities	Yes**	a,b
10. Development of large-scale agent manufacturing techniques	Yes	a,b
11. Target vulnerability studies	Yes	b
12. Selection of munitions/vector projects for research	Yes	a,b
13. Selection of munitions/vector projects for development	No	—
14. Standardization of weapons for possible procurement	No	—
15. Large-scale field testing with simulants, BW prototypes	No	—
16. Development of theory for battle planning	Yes	c
17. Preparation, voting of budgets for procurement and maintenance	No	—
18. Selection of standardized matériel	No	—
19. Large-scale manufacture, transport of agents	No	—
20. Manufacture, bulk transport/storage of munitions	No	—
21. Munition filling, storage	No	—
22. Deployment of matériel in forward areas	No	—
23. Training of individual troops in offensive techniques	Yes††	b
24. War games, including BW exercises	Yes††	b

* This table refers to activities during the 1980s. Prior to 1969 the United States maintained a large-scale offensive effort that incorporated each component listed here. The earlier offensive program would, in some important areas, obviate the need for extensive development and field testing in a present-day offensive effort.

† Stages adapted from SIPRI, *The Prevention of CBW: The Problem of Chemical and Biological Warfare,* Vol. V (1971).

** The DOD has not formally indicated that "development" activities are being conducted. But the pattern of work indicates all aspects of agent weaponization short of standardization and large-scale manufacture. This work is designated "defensive research" by the DOD.

†† Training for offensive use of *chemical* weapons, which is part of current U.S. regimen, is substantially the same as that of *biological* weapons.

a. 329 core projects.
b. Other DOD source materials, including obligation reports, other DOD reports and testimony to congressional committees, responses to lawsuits, projects indicated on official DOD lists of rDNA studies but for which no work unit summaries could be obtained, interviews with BW researchers and administrators.
c. U.S./NATO war-fighting strategy documents—see text for details.

This pattern of activities is particularly troubling in light of the inconsistency between the DOD's actual research activities and what Colonel Huxsoll, commander of the army's largest BW lab, calls the cornerstone of the medical defense strategy: broad-spectrum or generic vaccine development. Only 9 of the 329 core projects could reasonably be construed as focused on this goal. And available alternate sources mention broad-spectrum vaccines only in passing.

The simplest interpretations of this gulf between stated strategy and actual work are that the military either does not understand its own approach or is lying. Jonathan King, the biological warfare expert at MIT, suggests an additional explanation for this admixture of checkered quality and suspect intentions: The DOD is "casting a wide net," in order to recruit scientific talent and establish a flexible, broad base of technological support.

This is reflected in the fact that 43 of the 329 core projects—12 percent of the total—provided funding for graduate education fellowships, equipment purchases, and scientific conferences to discuss biotechnology problems of military interest. In many cases, the expectation that contractors would later take on military research was made explicit. For example, the army funded a University of Washington "Neurotoxin Research Facility" to "make possible or improve research which supports the Army thrusts in CBD (chemical and biological defense)." (The broad implications of this funding on the academic community are addressed in Chapter 9.)

"At this stage of the program it doesn't matter very much what the specific projects are or even if the work has any intrinsic scientific merit," King said. He cited a personal example: The Office of Naval Research (ONR) attempted to recruit King's department to conduct rDNA studies on marine biofouling—that is, how to use molecular biology to detach barnacles from ship bottoms. "Our department has no experience with or knowledge of marine organisms or environments. However, we are a world center of expertise in genetic engineering and molecular and cellular biology," he said. "The ONR's true goal is to form a reservoir of scientists to help the DOD understand and explore the technology's full potential. Under such circumstances it would hardly be surprising if much of the research were of mediocre or inferior quality."

Though this effort is described as defensive, of course, it is precisely how effective *offensive* efforts to exploit a new technology must begin. The research focus will narrow sharply when specific offensive military missions are clearly identified, King added. And its quality will markedly improve when weapons are developed and tested. That stage will be the polar opposite of the current, freewheeling program: a few precisely defined goals pursued with exacting, rigorous quality control.

In view of the range and depth of the military program and the futility of medical defense against today's BW threat, it would seem either that the Pentagon's scientists are largely incompetent or that their work is essentially offensive—albeit at an early stage of development. In all likelihood, each of these factors drives the military's work in BW and biotechnology.

THE DESERT HIDEAWAY AFFAIR

In October 1984 Tennessee Democratic Senator James Sasser revealed a plan that connects some of the dots in the Pentagon's arcane biological worldview. He unveiled an army proposal to build a $1.4 million biological weapons test laboratory in order to conduct secret research on "substantial volumes of toxic biological aerosol agents." The research would be defensive, like all its BW work, the army explained, and would be carried out under conditions of utmost safety.

The news sparked protests among a host of scientists. The following December Harvard molecular biologist Richard Goldstein spoke for many colleagues when he said, "In my mind, the opening of this facility substantially escalates the biological arms race."

"Whether or not one accepts the hypothesis that the new [lab] is offensive research wearing a defensive cloak," added Richard Novick, director of the New York Public Health Research Institute, "there is no question that it represents a major escalation in biowar activity."

The new lab would be the centerpiece of a planned major modernization and expansion of the Dugway Proving Ground in Utah. At the time these plans were announced Dugway was expected to add 309 employees to its payroll and double its CBW testing work load by 1988. "Field instrumentation, chemical life sciences, photographic, laboratory and environmental testing facilities will be upgraded," noted an army report.

In its request to Congress for funding, the army said the aerosol lab would be used "to evaluate biological defensive readiness and to test protective gear and detection/warning equipment by employing toxic microorganisms and biological toxins requiring a level of containment and safety not now available within the Department of Defense."

The lab would operate under P4 containment—the most secure physical protection possible—for experiments with some of the most dangerous pathogens known. Fort Detrick has six large superinsulated, negative air pressure P4 labs, but it tests only small quantities of pathogenic aerosols in its infectivity studies. The Dugway unit would be the only P4 lab in the nation devoted exclusively to nonmedical research.

The proposed tests would be of two general types, according to the army's request: "Biological defensive testing of military equipment and detection devices [and] evaluation of the foreign biothreat. The latter ranges from testing the vulnerability of protective masks or other equipment to assessing the characteristics of biological materials suspected of being used by potential adversaries in military or terrorist situations."

Although proposed appropriations with substantive policy ramifications are normally debated by Congress, the DOD buried the Dugway aerosol lab funding request in a routine application to transfer unspent funds from other army projects. Typically the army asks for "reprogramming" approval to complete such noncontroversial projects as parking lots or offices.

Reprogramming requests are usually rubber-stamped by the ranking Republican and Democrat on the House and Senate Appropriations committees' subcommittees on Military Construction. Sasser, ranking Democrat on the Senate subcommittee, initially approved the plan but later took the extraordinary step of withdrawing his assent.

In a letter to the subcommittee chair Mack Mattingly (Republican of Georgia), Sasser complained that there was no statutory authority for the project and that it raised "important questions with regard to the potential capabilities for testing and production of offensive lethal biological and toxin weapons." He concluded that the army sought "to avoid the regular authorization and appropriation process of the Congress."

At Mattingly's urging, however, the other subcommittee members fell into line and voted to override Sasser's objection. In a letter Secretary of Defense Caspar Weinberger tried to reassure Sasser. He said the lab would not be used to develop offensive BW and that the Defense Department did not intend to violate the 1972 treaty.

But many leading scientists questioned the need for this type of lab for purely defensive purposes. Prominent among these was MIT's David Baltimore, a Nobel laureate and one of the most influential molecular biologists in the world. "This is too elaborate a program," he said. It is "too open to ambiguous interpretation, even if [the army's] intentions are good."

The concerns of independent scientists focused initially on the army's rejection of biological simulants in favor of testing with highly dangerous pathogens. The army agreed that using simulants—the benign organisms that mimic the behavior of pathogens—would be a safer and therefore preferable method. But it said the results from simulant tests would be unreliable.

"The efficacy of simulants for various testing purposes cannot be determined without exhaustive comparisons between simulants and threat agents," one army report stated. "Establishing the adequacy of the simulant for a specific test may require more aerosol work with pathogens than the test itself." The army argued that calibrating ultrasensors and evaluating protective gear and decontamination techniques would therefore be impossible without use of the actual threat agents.

"Adequate data concerning persistence cannot be developed [by the use of simulants]," it continued. "Data such as infectivity, symptomology and lethality can also not be obtained with simulants, should these be required."

A few scientists outside the military, including Emmett Barkley, safety director at the National Institutes of Health, publicly supported this rationale. But a throng of leading independent scientists disagreed strongly. Harvard molecular biologist and CBW expert Meselson was among them.

Virtually all the toxins and BW agents to be tested in the aerosol lab can be simulated, he said, adding that the use of simulants is a better military strategy. "The characteristics of aerosols important for defensive work are particle size and surface tension, and we've known for a long time how to match these with simulants," he said. A single simulant organism could duplicate the characteristics of many viruses and bacteria. And it could pose a hardier, more generic test of protective gear.

"We know a good deal about the likely biological warfare agents, and one can easily choose nonpathogenic or avirulent agents with the same size and molecular properties," said Roy Curtiss, chairman of the biology department at Washington University. "I don't see the need for a P4 lab. It's overkill and it's not good science."

"Development of defensive and detection equipment can be performed with absolute confidence and . . . safety by the use of killed organisms," Richard Novick argued. "These will have precisely the same particle size and chemical composition as the live organism and no containment would be necessary; indeed, field conditions could be much better approximated by performing the tests outdoors. Decontamination analysis and tests of persistence do not require aerosols. Additionally, these tests can be performed in complete safety by the use of live attenuated organisms such as those used for vaccines."

Novick also pointed out that toxins, which are nonliving and nonpersistent, would not require P4 containment. And he questioned the army's need for a P4 aerosol lab to test infectivity, symptomology, and lethality. "Experiments with live human subjects would not be permissible," he said. "'Symptomology' is well known for most of the agents under consideration on the basis of natural disease and experiments performed by the Japanese during World War II on human subjects. 'Infectivity' and 'lethality' data . . . would be unnecessary for any imaginable purely defensive purpose."

Critics cited as well the inherent dangers of aerosolized pathogens, especially when used in "substantial volumes." Aerosols are "the most dangerous vehicle for dissemination and the most difficult to contain," Novick said. Absolute physical containment is "a theoretical as well as practical impossibility," he added. As few as ten organisms of the causative agents of tuleremia and Q fever—each planned for Dugway testing—can cause human infection. A single drop of such agents could contain billions of organisms. Even an extremely small chance of accidentally spreading a deadly disease might be considered excessively hazardous.

The DOD has repeatedly stated that a major goal of its CBW defense program is to verify that combat vehicles are leakproof and could effectively be decontaminated. Indeed, part of the planned expansion of Dugway is to test *chemical* exposure and decontamination in a facility large enough to operate tanks. But curiously, discussion about *biological* testing on the same scale has been conspicuously absent from military documents and statements, although the need for such tests is dictated by Pentagon logic.

"We continue to obtain new evidence that the Soviet Union has maintained its offensive biological warfare program and that it is exploring genetic engineering to expand their [*sic*] program's scope," Weinberger wrote to Sasser. "Our development efforts in this area are driven by the Soviet threat. To ensure that our protective systems work, we must challenge them with known or *suspected* Soviet agents [emphasis added]."

This statement is consistent with the DOD's long-standing public posture: defense against the Soviet threat. But in this context it also supports the contention that the testing of recombinant prototypes ("suspected Soviet agents") is the actual goal for a Dugway lab, which would be equipped with state-of-the-art facilities. "There is no question that this equipment will give the Army the capability to perform genetic manipulations. . . . If they actually begin such work, that would give me cause for concern," David Baltimore said. "They should also say that they are absolutely not going to make any new toxins."

"The infectious characteristics of most of the 'standard' agents are very well known," Novick said. "The filling of 'knowledge gaps,' suggested to be an important part of the program . . . arouses strong suspicion that new types of agents will be developed and tested for their potential as bioweapons."

DOD officials repeatedly stated that they had no plans to conduct rDNA work at Dugway and denied they would ever create prototypes. They refused, however, to rule out rDNA work completely or the use of Dugway facilities to test aerosolized recombinant microbes that had been engineered at other labs. It may be difficult for the Soviets to produce practical BW in the lab, Colonel Robert Orton, director of the army's nuclear-biological-chemical defense division, said. "But one surely has to look at all the possibilities and ensure that there isn't an easy way to do it."

Beyond the prototype question, genetic engineering work would present new safety-problems. The federal guidelines on rDNA research mandate absolute avoidance of aerosols. These guidelines also depend on "biological containment"—genetic crippling of potentially dangerous organisms to render out-of-lab survival impossible. Biological containment, of course, would be antithetical to the BW experiments suggested by the army.

The final point of contention on the Dugway proposal was secrecy. Secret work can only feed suspicions of impropriety, critics say. Because threat assessment and equipment vulnerability are normally kept secret, DOD spokespersons admitted that some of the research at the proposed lab inevitably would be classified. This could include genetic engineering work, they acknowledged, despite the DOD's self-imposed ban on secrecy in gene splicing.

After the Senate subcommittee had approved funding for the lab, author and genetic engineering critic Jeremy Rifkin and retired Navy Admiral Gene La Rocque, director of the Center for Defense Information, a liberal defense research group, filed suit to bar construction. They charged that the DOD had violated the National Environmental Policy Act by failing to prepare an environmental assessment to address the above safety concerns.

The suit spotlighted the proximity of Dugway to Salt Lake City (less than ninety miles) and the fact that Dugway also tests conventional weapons. This raises the concern that deadly microorganisms might be released should a errant projectile hit the lab. These fears gain credence upon review of army documents: The proposed lab site is less than five kilometers from a conventional weapons range, while Dugway's "artillery, missile, and mortar ranges may be used for firing up to 65 km." Such an accident seems unlikely. But the lax safety practices of Dugway testing programs during past decades, described in Chapter 2, give pause.

In January 1985, before the lawsuit was heard in court, the army agreed to file an environmental assessment. A cursory report was released in early February. Not surprisingly, the study found no significant environmental risks in the laboratory plan, largely reiterating earlier arguments. The army did make a surprising shift in its rationale, however.

The original reprogramming request to Congress justified the expenditure as "an urgent requirement . . . to provide an essential laboratory [that is] the only way to ensure [the] survival of our armed forces on the biological battlefield." This rhetoric was toned down substantially in the environmental assessment. The army concluded that it didn't really need a P4 lab for current work after all. It could make do with a more modest containment facility. Still, the new report stated, a maximum containment lab should be constructed in anticipation of some hypothetical future threat.

"The Soviets are exerting great efforts in this area. Consequently, we must develop appropriate defenses," the report noted. "Conceivably, such studies could involve laboratory operations which require the use of [P4] containment."

Rifkin and La Rocque immediately reactivated their suit on the ground that the assessment was inadequate. A hearing was held on April

26, 1986. The Pentagon made a concerted effort to persuade U.S. District Court Judge Joyce Hens Green that an aerosol test lab would pose no environmental or health threat. It further distanced itself from its original rationale for aerosol testing. No expansion of current testing was planned, DOD spokespersons now insisted, nor was there any clear idea of when a P4 lab would be needed. It was required, they said, "in anticipation of requirements which may never materialize."

This new stance generated the counterargument that a lab unjustified on emergent national security grounds, while stimulating substantial safety and arms control concerns, should not be built. Green sided with the plaintiffs. She slapped an injunction on lab construction pending an adequate environmental assessment.

In a related development in May 1986 the DOD released a response to questions by the House Committee on Appropriations about the Dugway lab proposal. The report is a surprisingly detailed and elaborate explanation of Pentagon views on biotechnology. Its basic message reverted to the DOD's earlier posture of urgency: The new genetic technologies combined with Soviet arrogance were depicted as leading to unprecedented dangers. Only a major U.S. biotechnology push to develop effective countermeasures can save the nation from a horrible fate, the report stated. And a Dugway aerosol lab, once again, became the key to survival.

This report made it clear that a legal setback had not dampened the Pentagon's resolve. But this time it is doing its homework carefully. By October 1987 the revised environmental assessment had still not been released.

A TACTICAL SHIFT

The large biotechnology program in general, and the Dugway lab affair in particular, seem provocative. But do they, in fact, betray a systematic U.S. retreat from offensive BW chastity? There is no certain answer to this question. But an important clue can be found in U.S. war-fighting doctrine, the theoretical framework that dictates military strategy and action in the field.

In August 1982 the army revised its field operations manual to incorporate two significant tactical changes prompted by new technologies. The first is the "extended battlefield"—the ability to make a "deep strike" behind enemy lines. New electronic technologies have endowed missiles with unprecedented accuracy. Distant targets can be hit with minimal "collateral" or unintended damage. The second change is the "integrated battlefield"—the ability to employ conventional, nuclear, or chemical weapons interchangeably as required by military needs and conditions.

Instead of offering the dichotomous stalemate of inaction or nuclear holocaust, the new strategies promote a victory-oriented posture. In December 1984 NATO began to adopt this extended battlefield concept, according to a leading West German military analyst, Alfred Mechtersheimer. "'Deep strike' has led unavoidably to a re-evaluation of chemical and biological weapons, which were previously neglected in military strategy."

Another aspect of the new offensive plans lends itself well to CBW. "The key factor is no longer destructive capability but rather the capability of rapid advance," Mechtersheimer has pointed out. "Ground-winning thrusts presuppose the use of captured weapons and infrastructure. This is a completely new task compared to [*sic*] previous NATO strategy, and along with neutron weapons ["enhanced radiation" weapons that are deadly but do not destroy physical surroundings], chemical and biological weapons are particularly well suited to this task."

CBW might even be considered superior to nuclear (including neutron) weapons, because it presents the requisite prospect of intermediate and deep saturation effects, but might be considered slightly less likely to provoke uncontrollable escalation. International law prohibits first use of CW. But in the heat of battle it may be impossible to determine which side used chemicals first.

In any case, a major army field manual makes the adoption of *chemical* weapons explicit. "Commanders must be prepared to integrate chemical weapons into nuclear and conventional fire plans on receipt of chemical release," it reads, and lists the advantages of chemicals "employed in mass and without warning."

Although NATO commanders are working on plans to integrate CW into their procedures, considerable conflicts about chemical deployment remain within the NATO alliance, as noted in Chapter 3. The changes in war-fighting strategy are a long way from becoming institutionalized.

But a doctrine that radically increases the utility of unconventional weapons, including BW, holds troubling implications. "If political and military leaders show no scruples in preparing for the use of atomic and chemical weapons of mass destruction, why should they dispense with biological weapons if they have military advantages," Mechtersheimer asked, "particularly since they can be produced cheaply, secretly, and on short notice?"

NATO and the WTO are silent on BW deployment, which is banned by international law. Bellicose speech making and devotion to BW research, however, demonstrate unwavering respect for the BW threat. Articles in specialty publications on both sides frequently treat biological and chemical arms as virtually the same. The term *chemical-biological warfare* has increasingly gained currency. An apparent shift in strategy alone, of course, is an ambiguous guide by which to judge U.S. motives and ambitions in the biological sciences. But it adds questions to the already burdened rationale of defensive research.

"NATIONAL INSECURITY"

Openness, say independent scientists, is the litmus test of defensive intentions in biological research. Secrecy—whatever the justification—breeds suspicion and skepticism on the part of domestic critics and enemies alike. Secret information undermines public involvement in policy debates because the national security apparatus can always lay claim to superior knowledge and reject opposing arguments out of hand.

In biological warfare the DOD is by no means an impenetrable monolith populated by sinister ideologues. To allay fears regarding genetic engineering, the department established in 1981 a policy that requires a complete record of each rDNA study to be kept on file at AMRIID. This was part of a multifaceted effort to "ensure that all DOD recombinant DNA research is thoroughly reviewed, properly approved, and can bear the scrutiny of an interested party," the policy reads.

Openness is a cherished value in the collegial milieu of AMRIID and other military labs, say the scientists who work there. Almost nothing in the medical defense program is classified, according to Colonel Huxsoll, AMRIID's commander.

"It's a whole lot easier to do the work if it's unclassified," Huxsoll added. "You do not have that barrier to acquisition of good research—whether it's in-house or a contract." At a recent international diplomatic conference secrecy was "an overwhelming issue," he continued, stressing efforts to keep any cause for suspicion out of his program. "I have a personal responsibility to see that that continues."

Although the AMRIID labs are within Fort Detrick, which, like all army installations, is reasonably secured, the institute is no fortress. Numerous technical exchanges among military, government, and university scientists from within the United States and from friendly nations attest to this fact. Journalists are given tours of the labs and support facilities.

AMRIID scientists, and their cohorts in other military labs conducting research into medical defense against BW, are free to publish the results of their studies in the scientific literature. The army even follows Food and Drug Administration rules and procedures in licensing their vaccines and antiviral drugs, Huxsoll said. And these products are often shared with public health authorities around the world.

Compared with the extreme secrecy in which many military programs are shrouded, Huxsoll's operation may seem to be operating in a fishbowl. But a number of actions, discrepancies, and policies cloud the water and pose serious doubts about the credibility of DOD claims that its medical defense program is completely unclassified.

The "interested party" referred to in the DOD's openness policy, we've concluded after years of investigative research, has to be possessed of an interest akin to relentless determination in order to check up on the military.

It is a simple matter to obtain from Fort Detrick a list naming its rDNA studies, investigators, and their institutions. Information beyond single-phrase descriptions, however, requires a Freedom of Information Act (FOIA) request. The DOD routinely delays responses to such requests for months beyond the ten-day statutory deadline. The process can involve numerous letters, phone calls, and a considerable investment of time even on relatively straightforward petitions.

In analyzing the responses to many FOIA requests filed over a period of four years, we found none of the official lists of rDNA studies to be comprehensive. The lists we obtained in spring and fall 1983 and fall 1986 either were outdated or omitted certain studies. Documents released to us by a wide range of other military agencies prove these discrepancies unequivocally.

No similar list is known to exist for biotechnology work not involving rDNA. So when we sought biotechnology work unit summaries under the FOIA, we made identical requests to several military branches. Because each branch relies on the same central computer system, we could cross-check the responses for accuracy and completeness.

The result was far from reassuring, with major differences between the responses. And as noted in the above analysis of the 329 core projects received from these requests, portions of the biotechnology program—clearly identified in other unclassified sources—were not represented in the work unit summaries released to us. These included substantial development work in industrial-scale bioprocess technology and major microencapsulation, aerosol vaccination, and rDNA research projects.

It is difficult to know if such "errors" are due to a failure of administrative control, or due to a lack of understanding of the technologies, or are the product of deliberate subterfuge. These are perhaps equally disturbing possibilities.

It may also be misleading to take comfort in the fact that military scientists are allowed to publish in academic journals. CIA researchers who conducted studies on the potential of LSD in mind control for the MKULTRA program during the 1950s and 1960s also published their work. "But those long, scholarly reports often gave an incomplete picture of the research," according to *The Search for the "Manchurian Candidate,"* the most comprehensive volume on these mind control activities, written by John Marks.

"In effect, the scientists would write openly about how LSD affects a patient's pulse rate, but they would only tell the CIA how the drug could be used to ruin that patient's marriage or memory," Marks commented. "Those researchers who were aware of the Agency's sponsorship seldom published anything remotely connected to the instrumental and rather unpleasant questions the MKULTRA men posed for investigation."

Harold Abramson, a New York allergist and psychedelics pioneer, was a case in point, according to Marks. "Abramson documented all sorts of experiments on topics like the effects of LSD on Siamese fighting fish and snails, but he never wrote a word about one of his early LSD assignments." A 1953 document indicates that the CIA charged Abramson to test LSD for its ability to elicit aberrant behavior, alter sex patterns, or create dependence.

This is not to say that AMRIID operates in the same fashion. But this much is clear: The right to publish per se is an insufficient test of openness. Indeed, many of the military's primary BW researchers do not publish at rates even remotely consistent with the vast resources at their disposal.

The commercial data base MEDLINE, compiled by the National Library of Medicine, is the most comprehensive on-line source for worldwide biomedical literature. We tapped MEDLINE for records going back to 1981 on nineteen leading investigators from various DOD units, as indicated on the 329 work unit summaries.

The data were startling: Seven scientists published only in obscure, poorly distributed, undistinguished journals, such as the *Southeast Asian Journal of Tropical Medicine and Public Health* and the *Scandinavian Journal of Infectious Disease.* Four of these investigators, who collectively spent more than $7.8 million dollars from 1982 to 1986, published a total of only seven papers. At $1.12 million per article, these may be among the most wasteful biological research projects in history. Either that, or the touted "open publication" policy is a sham. (The larger implications of this record are discussed in Chapter 9.)

A QUESTION OF DEFINITION

Beyond medical defense, much in the BW research program is classified. This includes some work on ultrasensors and protective equipment, assessment of suspected threat agents (such as "yellow rain"), and intelligence information. The problem lies in definition and dual-use research. Just as the parameters of "medical defense" are vague, so the technologies used are applicable in many areas.

MCA techniques, for example, form the backbone of ultrasensor development. The same methods are the key to creating vaccines. Knowledge gained from examining novel biothreats, which might be classified, has direct vaccine applications. Of course, it may have equally plausible offensive uses, such as precisely targeted toxin weapons.

Most scientists in AMRIID and other military labs conducting unclassified research must obtain security clearance. This requirement has nothing to do with their work, according to Huxsoll. It merely allows them to see classified intelligence information. Again, the distinctions blur. Do ultrasensor researchers also have access to classified data involving the experimental analysis of novel threat agents? If so, how does this influence their own work?

The DOD views biotechnology expansively, applying it to everything from barnacle removal to chemical warfare. Marine biofouling studies probably contain little pressing national security data. But this may not be the case with chemical warfare.

The CW program is largely classified and makes liberal use of biotechnology for antidotes and weapons development as well as to gain a better general understanding of chemical effects at the molecular level. By its nature, such research has many applications. What is to prevent the army from using properly classified CW research in its BW program?

The Pentagon's efforts to allay such concerns are often less than soothing. "You have to draw a distinction between a [BW] program—a generalized program. We don't classify any of that," said nuclear-biological-chemical defense chief Orton in an interview. "It's when you get into something that might identify a specific vulnerability of a member of the U.S. forces or of the U.S. forces in larger scale to something a potential enemy might use; then we would classify it."

Translation: Any militarily meaningful advance can be classified at army discretion. It is difficult, if not impossible, to identify precisely the programmatic distinction between "secret" and "open."

Ambiguity in secrecy criteria aside, the DOD's logic in classifying *anything* in the BW program tends to belie stated defensive intentions. Ultrasensor systems must be kept secret, the DOD says, to prevent the Soviets from developing ways to defeat them. "Unfortunately, on the one hand, a comprehensive system for the detection and identification of all possible agents is unimaginable," said Novick. "On the other hand, the principles of general detection are obvious enough and a reasonably sensitive system cannot be circumvented."

There is no reason to classify any detection or protective measures, according to Meselson. "It's only through force of habit that they're doing it. With the exception of intelligence work, we've got to be able to say, 'Absolutely everything we're doing, we're doing out in the open.' We must get rid of secrecy. Secrecy is the real threat."

The extent of military illogic in classification was apparent in a 1986 interview with Huxsoll. He freely acknowledged that AMRIID holds, in small quantities suitable for research, samples of all the agents in the pre-1969 offensive program as well as suspected threat agents discovered since then. This includes dozens of deadly toxins and hundreds of strains of viruses and bacteria. He identified scores of these agents by name in a recorded interview and gave parameters for the maximum amounts maintained at any time.

But Huxsoll would not release a complete list with precise quantities. The reason: terrorism. He suggested a terrorist or psychotic might see the list and want to break into a fortified U.S. Army base to steal some BW agents. Presumably only the official list with definite amounts would tempt a terrorist.

Commercial factors also contribute to medical defense secrecy. AMRIID sponsors a major "Cooperative Antiviral Drug Development Program," in which private companies work with the army to develop products that might have both BW defense and civilian applications. Included in the contracts are explicit, broad protections for "proprietary information." AMRIID routinely agrees not to publish or otherwise not to disclose test data or any other aspect of the studies.

Of course, proprietary information is routine in corporate America. But it is still *secret.* Asked if this constitutes an intrusion of secrecy into an otherwise open program, Huxsoll insisted that such arrangements are essential to BW defense because joint projects are highly cost-effective and companies must be able to protect their investments. Without denying the existence of this kind of secrecy, he dismissed concern over it as naive.

Others are not so sanguine. "Such relationships are highly susceptible to abuse," said King. "Corporations—particularly small, struggling firms—might recognize tremendous economic incentives to conspire or collude with the army about how much information to release and when to release it. Protecting proprietary information could act as a cover for classifying defense efforts."

CUTTING OFF THE SOURCE?

In 1984 the Reagan administration reasserted the importance of controlling the flow of biotechnology know-how to the Soviets and began to press for stringent barriers on the export of what Defense Intelligence Agency spokesman John H. Birkner called "keystone equipment."

This includes high-containment facilities, large-scale fermentors for growing biological organisms, ultracentrifuges, and other equipment that could be used to refine toxic biological products. These were ostensibly added to prohibitions on the export of certain organisms and toxins that appear on the Pentagon's "Militarily Critical Technologies List."

Corporations that might be tempted to export such devices would confront government officials who are charged with "rooting out all those who would cooperate with the enemy," Birkner warned. "In such a contest," he added ominously, "government would probably prevail."

The problem with such controls is that the international biotechnology equipment trade is already widespread. The tools of biotechnology are accessible to any reasonably advanced nation through domestic production or trade with European countries.

Such an embargo, therefore, would have little impact on the Soviets. It would instead "markedly harm our country's ability to compete in overseas markets," J. Leslie Glick of the Genex Corporation told a House of Representatives subcommittee. More important, Birkner's approach could have a devastating impact on the free international exchange that nurtures the development of what is, in essence, a basic-science industry. In the meantime, arms control may actually be hampered by such restrictions if other nations interpret them as evidence of heightening U.S. respect for the value of genetically altered BW.

Information wars soon eclipsed the export debate. In May 1985 the Defense Department released a new publication for military contractors seeking access to technical data in order to prepare bids. The image on the cover of the document is illuminating: a hammer and sickle with a bold line through it. "Stopping the Soviets' extensive acquisition of military-related Western technology in ways that are both effective and appropriate in our open society is one of the most complex and urgent issues facing the free world today," according to the booklet's opening statement.

The document was prepared for the Defense Technical Information Center (DTIC), the main computer conduit for public access to defense-related information. Since 1985 restrictions have been placed on even *unclassified* information within the DTIC. For example, the work unit summaries we used for our analysis, once distributed relatively freely from DTIC files, are now restricted to qualified government agencies and contractors or are released only reluctantly and incompletely to reporters willing to assert their rights under the FOIA.

In October 1986 National Security Adviser John M. Poindexter elevated the practice to the level of official policy. He expanded its application to the National Technical Information Service (NTIS), largest unclassified U.S. data bank, and to commercial computer networks. It was one of his last acts before he resigned in the face of the Iran-contra arms scandal.

The logic of the policy was that by sifting and organizing a multitude of unclassified documents, enemies or economic rivals could construct a "mosaic" that would reveal information damaging to U.S. interests. Sophisticated public access communications systems made this a growing danger, according to Poindexter.

The DOD moved quickly to constrict the flow of information

deemed "sensitive but unclassified." The Pentagon itself defined the parameters of "sensitive." (One military official admitted that the definition was so broad that "it covers anything anyone wants it to.") The new policy offered major opportunities for erecting arbitrary barriers against public access to information needed for oversight of military activities, including those in biotechnology.

The Pentagon immediately announced its intention to implement curbs on access to such nongovernmental, commercial data banks as Nexis—a service operated by Mead Data Central, Inc., that catalogs articles from a wide range of newspapers, magazines, and unclassified government sources.

Executives of commercial services and civil libertarians, not surprisingly, were outraged. They questioned the legality of this kind of Pentagon intrusion. But as far as the DOD was concerned, the case was closed. "The question is not 'will there be restrictions or controls on the use of commercially available on-line data?' The question is 'how will such restrictions or controls be applied?'" said a DOD spokesperson in late 1986. "We are very serious about protecting information, including unclassified but sensitive information."

In the wake of the Iran-contra scandal revelations of secret policies, however, congressional animosity toward this new executive branch initiative peaked. In response to widespread attacks the Reagan administration withdrew the Poindexter policy in March 1987. As a test of the Pentagon's "big brother" aspirations, however, the episode was chilling. And in light of the multitude of loopholes in the DOD's "openness" about biological warfare and biotechnology, the true depth and range of its program may vastly exceed what the public has been allowed to discover.

DECODING MILITARY SPENDING

Beyond program and research descriptions, the military budget provides the final insight into the breadth and direction of the BW effort. It is also a highly complex puzzle. Just as the overlap in DOD programs makes it impossible to know how much biological warfare research is kept secret, so budget categories also overlap. Medical research in the *chemical* warfare budget, protective equipment development, and general biomedical and environmental research all may include biotechnology studies that could be used to solve problems relevant to *biological* warfare. In this sense the designation "BW defense" for some studies, but not others, is arbitrary—at least to outside observers.

And because the range of DOD biological research indicates an infrastructure of talent and expertise that identifies with the military and relies on Pentagon support, it would be misleading to focus solely on the "biological defense research" budget to the exclusion of other DOD spending in the life sciences.

The term *biotechnology* has no uniformly accepted definition. Funds spent on "medical defense against biological warfare" include everything from classical biochemical procedures established decades ago to the newest, most sophisticated genetic manipulations. MCA and rDNA technologies are unmistakably part of the bioengineering revolution. But many methods cannot be discretely separated into "classical" and "novel." They lie on a continuum, which constantly changes with the field's rapid innovation. For this reason we present overall research figures, although our analysis emphasizes the specific impacts of the new biotechnologies.

Again, interpretation of the data must be tempered. The Pentagon says its entire BW-related budget is unclassified, as mandated by federal law. But the secrecy inherent in military operations and the overlap in research categories makes it impossible to evaluate that claim with certainty. Each year the DOD budget contains a classified portion known as the "black budget" because it is blacked out in published material and is not publicly accountable. Even Congress may have little information on this spending.

Black-budget requests have tripled during the Reagan years—to $35 billion in the fiscal 1988 request—according to the *Philadelphia Inquirer*. Now at 11 percent of the total, secret spending is growing faster than any other defense category. A variety of other techniques is used to obscure secret expenditures, the *New York Times* reports. Sometimes funds are listed by code words. In 1986, for example, the DOD "sought $10.9 million for a program labeled in public documents as 'Grass Blade,'" with no other description. Funds for secret projects may also be hidden in unclassified programs, the *Times* noted.

Therefore, although the figures we present are the best *public* estimate of DOD spending, *actual* spending levels are probably far higher. Table 10 summarizes the Pentagon's official accounting of its BW budget.

The BW budget has grown rapidly during the 1980s—a drastic departure from the Carter years, when the program languished. The average annual increase during the first five years of the Reagan admin-

TABLE 11

DOD RESEARCH OBLIGATIONS ADJUSTED TO REFLECT AVERAGE INCREASES FOR ALL FEDERAL AGENCIES ($MILLIONS—CONSTANT 1982)

FISCAL YEAR	ACTUAL DOD FUNDING	AVERAGE FEDERAL CHANGE FROM PREVIOUS YEAR	AMOUNT DOD WOULD HAVE RECEIVED IF INCREASED AT AVERAGE FOR ALL AGENCIES*	FUNDS ALLOCATED TO DOD ABOVE AVERAGE FOR ALL AGENCIES
ALL LIFE SCIENCES FUNDING				
1980	176			
1981	185	−3.8	169	16
1982	209	− .3	168	41
1983	214	4.7	176	38
1984	227	4.8	184	43
1985	243	9.2	201	42
			Net shift—1981–85:	180
MEDICAL SCIENCES FUNDING†				
1980	98			
1981	105	−4.8	93	12
1982	119	2.7	96	23
1983	129	4.3	100	29
1984	135	5.5	106	29
1985	152	9.8	116	36
			Net shift—1981–85:	129

*This represents a hypothetical calculation that applies the average increase in funding for all federal agencies, to DOD funding, using 1980 as the base year (see text for details).
†Subset of life sciences.

Source: National Science Foundation, "Federal Funds for Research and Development," fiscal years indicated.

To construct the table, we began with the DOD budget for life sciences and medical sciences (a subset of life sciences). The actual DOD figures, presented in the left column, reflect a far higher rate of increase than that which the federal budget as a whole experienced in these categories. The average annual increase for DOD life sciences research from

1981 to 1985, for example, was about 7 percent, after inflation. Overall federal expenditures in this category rose an average of less than 3 percent.

Starting with the last year of the Carter administration, we adjusted DOD increases for each succeeding year as if they had received the average federal increase for life sciences. By subtracting the *revised* figure from the *actual* DOD figure, we were able to calculate the net shift of funds from civilian to military research.

The results are sobering. Over a period of five years the Reagan administration shifted $180 million from civilian to military biological research. In medical sciences research, the shift—$129 million—was even more dramatic in relation to the total federal funding (see Figure 2). During this period the federal government spent an average of more than $5 billion a year on life sciences research. In this context, $180 million is hardly an overwhelming shift. But it is a drastic departure from past practice.

FIGURE 2

ACTUAL AND HYPOTHETICAL DOD OBLIGATIONS FOR LIFE SCIENCES AND MEDICAL SCIENCES RESEARCH

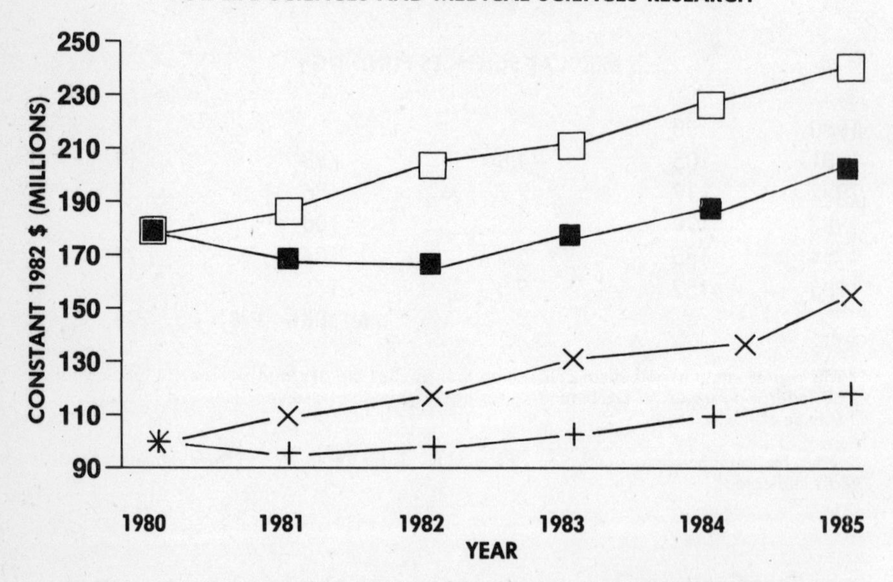

□ Actual DOD obligations, life sciences research ■ DOD obligations, if calculated at total federal rate of change × Actual DOD obligations, medical sciences research (subset of life sciences) + DOD obligations, if calculated at total federal rate of change

Source: National Science Foundation.

"I wouldn't be in this field if I hadn't gotten money from the Army," Brigham Young University biochemist Donald Robertson, who is conducting rDNA experiments on anthrax bacteria, told the *Wall Street Journal*.

"A bunch of my colleagues who 10 years ago would have looked down their noses at the Defense Department," added Francis Hoskins, a DOD contractor at the Illinois Institute of Technology, "are coming to me now asking how they can get their grant approved." Hoskins's genetic engineering work concerns nerve gas antidotes.

In December 1985 Colonel Franklin H. Top, director of the Walter Reed Army Institute of Research in Washington, D.C., warned scientists in his biochemistry division that they would soon be out of jobs if they didn't form "a credible research group in biological toxin defense."

These kinds of imperatives are clearly having an impact on the conduct of biological research in the United States. The long-term effects on academic scientists and on the American BW program are addressed in Chapter 9.

MAPPING INTENTIONS

What can be predicted from this funding trend? The carte blanche enjoyed by the Reagan administration in its early years has ended. But by any measure—budgetary, ideological, strategic, or political—the exploitation of biotechnology has been institutionalized in the Defense Department. Pressure to expand and modernize the BW program will survive in some form no matter who runs the government.

As noted in earlier chapters, the private sector has wielded considerable clout in CBW development and policy-making. The role of the chemical industry, which stood to lose hefty profits, in scuttling U.S. ratification of the Geneva Protocol is perhaps the classic case.

A more recent example of the compelling commercial incentives in boosting CBW is Agent Orange. In 1978 a class-action suit was filed on behalf of Vietnam veterans who suffered severe health effects from exposure to dioxin, a contaminant of the herbicide sprayed in Vietnam by the U.S. Army through the early 1970s. This led to disturbing revelations about the chemical's manufacturers.

Massive documentation from the suit shows that "Dow, the main supplier of Agent Orange, and other chemical companies knew the toxic dangers of dioxin exposure as early as the mid-1960s," according to a report by the Southeast Asia Resource Center, a Berkeley, California, think tank. "But the companies continued to sell dioxin-contaminated Agent Orange. In one 1967 memo, Dow scientists stated that chloracne, a skin condition related to dioxin exposure, is 'usually not disabling, but may be fatal.' In 1965, the Dow director of toxicology wrote that dioxin-caused 'fatalities have been reported in the literature.'"

The companies may have understood these dangers even as early as the 1950s, the report added. Ironically they based part of their legal defense on arguments that the government knew as much about the dangers of dioxin-tainted Agent Orange as industry did.

In recent years the chemical industry throughout the developed world has been tethered by growing export restrictions in response to evidence that uncontrolled international trade has contributed to CW proliferation. The Iraqis, for example, are believed to have established their chemical capability using equipment and CW precursors supplied by European firms.

The stigma associated with the manufacture of Agent Orange and napalm, along with concerns over proliferation, have discouraged some large companies from seeking military contracts for CW. This has combined with the recognition that although nerve gas production has once again been authorized, it is an unstable business, subject to curtailment by a recalcitrant Congress. As a result, when the army requested bids for binary nerve gas production in 1985, the largest companies were not interested.

DOD officials also fear that CBW defense matériel "might never offer a market large enough to attract the industries that could produce it," according to a 1985 congressional General Accounting Office report. "The problem might be particularly acute for medical equipment, because chemical and biological defense drugs would comprise [*sic*] a very small share of a pharmaceutical company's business (Army medical officers place the estimate at less than 1 percent)," the report continued.

"Chemical and biological defense drugs are limited production items; DOD might place an order for two production runs followed by none for five years," the GAO noted. Vaccines present similar headaches. "Making firms interested in the diseases of biological warfare is considered difficult, because these diseases are extremely rare in peacetime, making DOD the only customer," said the report. In light of the long lead time and the cost of compliance with elaborate federal regulations associated with drug and vaccine development, many firms reject DOD contracts as unlikely to produce an acceptable return on investment.

"For biotechnology applications such as decontamination, development officials say the incentive for major firms is nonexistent," the report added. "Chemical and biological defense is simply too small an area compared to [*sic*] the rest of the commercial potential for biotechnology, and developments outside chemical and biological defense do not transfer easily."

These factors have opened the door to smaller entrepreneurial firms, which are plentiful in the young biotechnology industry. Small firms have more compelling immediate needs for capital and may be less concerned with market potential years down the line than would larger, diversified corporations.

Venture capital is still flowing freely into hundreds of start-up companies. But most industry observers agree that a shakeout is inevitable. Some small companies will fail. They will not achieve the breakthroughs that lead to successful products in time to satisfy impatient investors, or they will not be able to compete effectively. When this happens, the increases in biotechnology funding from the Pentagon may assume ominous implications that dwarf the dollar amounts involved.

"Struggling biotechnology firms will increasingly turn to the DOD for a bailout if the money is there," King has predicted. "Appeals to patriotism and economic desperation combine to create a compelling incentive to accept DOD contracts. But private companies, because of their relative independence from oversight and peer review, may be more willing to bend rules and overlook potentially offensive or illegal research." This phenomenon (discussed in Chapter 8) is already apparent in the failure of several corporate rDNA contractors to register safety committees with the NIH, although DOD rules require such registration.

Military control of biotechnology is on the rise. Each year it becomes more integral to Pentagon BW planning and more integrated into its academic and commercial network. As will be shown in the following chapters, a parallel oversight process is conspicuously absent. In the face of secrecy, offensive research implications, vast expenditures, and vague doctrines that defy logic, self-policing in biological warfare research is a perilous way to keep the arms race in check.

7

THE 1972 BIOLOGICAL WEAPONS CONVENTION—OBSOLESCENT DIPLOMACY?

A HISTORIC ACTION

The Convention on the Prohibition of the Development, Production and Stockpiling of Bacteriological (Biological) and Toxin Weapons and on Their Destruction, signed in 1972, represents the farthest-reaching multilateral arms control treaty ever, mandating destruction of an entire class of weapons. The BW Convention stands as testimony to the near-universal revulsion people feel for these weapons and to the determination that they be consigned forever to history.

Nations that ratified the BW Convention agreed into perpetuity "never in any circumstances to develop, produce, stockpile or otherwise acquire or retain: (1) Microbial or other biological agents, or toxins whatever their origin or method of production, of types and in quantities that have no justification for prophylactic, protective or other peaceful purposes; (2) Weapons, equipment or means of delivery designed to use such agents or toxins for hostile purposes or in armed conflict." (See Appendix 3 for text and implementation status.)

Ratifying nations pledged to destroy existing arsenals of such agents and not to transfer biological or toxin weapons or technology to "any recipient whatsoever." Each party agreed to implement the treaty into domestic law through its own constitutional process.

In addition, the convention established a complaint and investigation procedure through the UN Security Council. Mindful of the erosion of past arms agreements by new technologies, the treaty mandated a review conference to take place five years later in order to "take into account any new scientific and technological developments relevant to the Convention." A further provision allowed a government to withdraw from the treaty after three months' notice "if it decides that extraordinary events . . . have jeopardized the supreme interests of its country."

At a minimum the BW Convention makes large-scale production and deployment of BW or TW vastly more difficult by driving these activities deep underground. They would have to be conducted in extreme isolation and at the risk of grave international repercussions. Thus, the treaty stands as a unique monument to restraint. Once a nation has adopted a policy to follow the treaty, that policy could not be reversed short of elaborate preparations. These weapons could never be used impulsively in a moment of crisis—not even by an insane aggressor. In contrast, no matter what the restrictions on numbers or types or deployment of nuclear missiles, no matter how careful and belabored their launch procedures, the decision to start a nuclear war takes only an instant.

By the end of 1985, 103 nations, including all the major powers, had become parties to the BW Convention. By at least one important measure the pact has been an unquestionable success: No nation has verifiably violated its treaty obligations.

In addition to the 1925 Geneva Protocol and the BW Convention, two other treaties have some bearing on CBW. The 1948 Genocide Convention forbids "acts committed with intent to destroy, in whole or in part, a national, ethnical, racial or religious group." This treaty effectively outlaws ethnic weapons (as described in Chapter 5) whatever their means of production. By the end of 1985 ninety-five nations had ratified the pact, signed in the wake of Nazi atrocities; they included most world powers, although some of these joined only in the past few years. The United States ratified in 1986.

The 1977 Environmental Modification Convention may also apply to BW. It prohibits any environmental manipulation, such as large-scale introduction of nonnative BW agents that displace indigenous microbes. This treaty is supported by only fifty nations, but the United States and the Soviet Union are among them.

A CHANGING GEOPOLITICAL PICTURE

A fortuitous combination of factors led to the BW accord. "By 1972 many in the military had become skeptical of the practical utility of BW," MIT's Jonathan King has pointed out. "It was seen correctly as a weapon that is difficult or impossible to control, likely to affect civilian populations more than armies, and of minimal strategic importance in the age of nuclear weapons."

Unlike the case of nuclear and chemical weapons, there was no civilian lobby to exert pressure against the treaty in order to promote and exploit Pentagon interest. There was little money to be made in the BW business. Perhaps most important, Richard Falk, Princeton University professor of international law, has pointed out, "The early 1970s were the highwater mark of détente." Arms agreements enjoyed broad support.

By the 1980s, however, each of these conditions had changed substantially. The advent of rDNA technology stimulated serious reevaluation of the potential for developing practical BW. The biotechnology industry was born and began to assert its economic interests by downplaying the potential dangers of genetic engineering. (The role played by industry in the BW controversy is described in Chapter 8.)

"A gradual return to the cold war had ensued, a dramatic quickening of the arms race had occurred, and considerable scepticism had been expressed, especially in the United States, with respect to the viability of arms control as an ingredient of national security policy," stated Falk. This geopolitical backdrop, he added, "tends to breed recrimination, propaganda and counter-propaganda campaigns, and an attitude of contempt for law-oriented approaches to international problem solving."

Under these conditions, the sometimes vague wording of the BW Convention has become a cause for concern. It should be understood that the treaty language was not a product of ineptitude. Like the Geneva Protocol and other multilateral arms control agreements, the convention was written broadly in order to garner the widest support; the wording indicates just how much the parties were willing to restrict themselves. After recent political and scientific developments, however, even strong supporters of the treaty regime have acknowledged that some of its broad language adds up to weaknesses that resemble gaping loopholes.

One of these weaknesses is that the treaty delineates no effective verification measures. To be sure, this is a tremendously difficult proposition in BW. Unlike other weapons systems, national technical means of verification—sophisticated satellite and radar networks—by themselves are insufficient to monitor BW violations. New bioprocess technologies have exacerbated the problem by reducing the size—and thus visibility—of production plants needed to produce militarily significant quantities of biological agents.

Complaints and investigations are handled by the UN Security Council—an inadequate forum at best. Any of the Council's five permanent members, which include the two superpowers, can veto any action. And the treaty has no provision for enforcement in the event that a violation occurs.

The treaty prohibits development except for "prophylactic, protective or other peaceful purposes." This wording is understood to ban deterrent arsenals. But the term *development* is not defined, with no consensus on where to place the dividing line between prohibited and permitted research and development.

As noted in Chapter 5, the U.S. military acknowledges that the difference between offensive and defensive BW research lies almost purely in intent. For example, the development of genetically engineered prototype agents could be viewed as a legitimate defensive inquiry under the terms of the treaty. Moreover, the treaty permits possession of BW agents and toxins for defensive research but fails to set limits on quantity. The extreme toxicity or virulence of some agents renders them effective even in minute amounts.

Asked in an interview for his interpretation, Colonel David Huxsoll, commander of BW medical defense at Fort Detrick, indulged in a bit of gallows humor. "We might have enough to kill *you* with some of these things. But that's not a weapon," he responded. Huxsoll declined to specify what the army considers an upward limit but said three liters of a given agent are often grown to prepare proteins for vaccine research. "If you had an eighteen-wheeler [truck] with a big tank on it, if we had enough to fill up one of those [tanks] with a given agent, that's too much," he added.

Another apparent loophole, involving the definition of *toxin,* was opened by rDNA technology. Until the mid-1970s toxins had been defined as nonliving poisons produced naturally by living organisms, including a wide range of bacteria, fungi, plants, and animals. But advances in both rDNA and protein chemistry have created the prospect not only of more effective TW but of "synthetic toxins."

Lethal fragments of toxin molecules normally produced by rare or fragile organisms may be cloned into hardy strains of bacteria. Hybrid toxins combining such fragments in new variations, exact replicas of naturally occurring toxins, and entirely novel toxins can in principle now be created and mass-produced.

It may seem obtuse to question whether such substances are covered by the BW treaty, which clearly bans toxins "whatever their origin or method of production." But a toxin fragment or synthetic toxin might be defined as a chemical. This would eliminate it from any legal stricture against possession and might quickly lead to increased toxin research as well as stockpiling for deterrence or a first strike.

A QUESTION OF TACTICS

In 1980 the first BW Convention review conference was held. The conferees affirmed that the treaty "has proved sufficiently comprehensive to have covered recent scientific and technological developments." Further, they said new agents developed using rDNA would be "unlikely to improve upon known agents to the extent of providing compelling advantages for illegal production or military use in the foreseeable future." In essence, this view holds that changes in BW technology have been quantitative rather than qualitative; therefore, fundamental changes in existing sanctions are unnecessary.

But the accelerating pace of biotechnology continued to generate fears—in and out of government—that the treaty regime was deteriorating and that the prospect of biological warfare was on the rise. Manfred Hamm, an analyst for the Heritage Foundation, a conservative think tank, echoed the views of many in the Reagan administration when he said the treaty "amounts to little more than an expression of universal aspiration."

Nongovernment scientists and arms control experts generally agree that even if the new biotechnologies were fully exploited, it would be extremely difficult—perhaps impossible—to create truly controllable

BW agents. And they remain skeptical about the utility of weapons that are unpredictable or only moderately predictable in their spread and efficacy. But this same community of experts is divided about the impact of biotechnologies on the treaty regime and about how to interpret these questions for both political institutions and the general public.

One position is closely aligned with the Pugwash movement, an influential private arms control research and advisory group. ("Pugwash" was derived from the name of Cleveland financier Cyrus Eaton's summer home in Nova Scotia, where Eaton sponsored "unofficial" meetings of Soviet and U.S. arms control experts in the 1950s.) Pugwash counts among its members many leading figures in CBW arms control, including Julian Perry Robinson of the University of Sussex, England; Jorma Miettinen, director of Finland's Project for Chemical Disarmament; Robert Mikulak of the U.S. Arms Control and Disarmament Agency; Harvard's Matthew Meselson, who is credited with strongly influencing President Nixon to renounce BW; and several Soviet scientists.

Pugwash endorses the 1980 review conference stance. Fears about the effects of genetic engineering on BW development and the treaty are "largely misplaced," the group believes, "since the potentials for misuse of that technology appear to be no greater than standard microbiological techniques which have existed since the inception of the 1972 Convention."

Without ignoring the reality of international concern about genetic engineering, Pugwash sees confidence building and enhanced openness and communication between nations as the way to keep the lid on the BW race. Pugwash Secretary-General Martin Kaplan went so far as to suggest that even encouraging public debate about the potential weapons applications of biotechnology is dangerous because it could stimulate military interest or help legitimize increased research by both East and West as well as spur moves to limit civilian biotechnology.

Another position has been loosely developed by a range of groups and individuals, notably the Green party of West Germany and the Committee for Responsible Genetics (CRG), a Boston-based public interest group. The CRG board includes MIT's King; former members of the NIH Recombinant DNA Advisory Committee Sheldon Krimsky of Tufts and Richard Novick of the New York Public Health Research Institute; Nobel laureate in biology George Wald of Harvard; Victor Sidel, former president of the American Public Health Association; and a variety of other leading thinkers in the social and technical implications of health and science.

These experts agree with Pugwash that building confidence and openness are essential. But they also think—as most parties to the BW Convention have come to agree—that advances in biology pose new threats to biological arms control. These groups call for greater control over military research, including a ban on secret research and restricting the range of permitted studies. The Greens have demanded a moratorium on all military research using genetic engineering.

The dangers of broad, substantive public discussion of these issues are far outweighed by the contribution of open discourse to popular control over military initiatives, according to the CRG and others. They argue that attempts to shield the public for fear of encouraging the military would merely ensure a low quality of debate and promote control by military and government technocrats. To an extent, these fears have already been realized. In the United States particularly, BW developments are poorly understood by the general public, scientists and legislators alike.

In the meantime, however, U.S. military leaders are unified around the explicit view that genetic engineering has profoundly, irrevocably altered the BW landscape. While scientists weigh the exigencies of military biology, the Pentagon has built a massive BW defense apparatus. Its size and nature are lightning rods for speculation and uncertainty about the U.S. commitment to biological arms control. The Dugway aerosol lab proposal is a particularly cogent example.

"Why should we accept the self-interested assurance of the Army that it is not engaging in prohibited types of research at Dugway?" Francis Boyle, professor of law at the University of Illinois, has asked. "If the Soviets were to make this same type of assurance for a similar facility that they built for allegedly defensive research on biological weapons, we would not believe them. And rightly so." Under current conditions of distrust, Boyle added, such provocative actions could begin "an action-reaction cycle . . . and both sides will in fact quickly be acting as if the treaty were a nullity."

Another question about U.S. intentions involves its failure to enact specific domestic legislation that imposes penalties on BW development by corporations, private citizens, or universities. This raises the prospect that a struggling biotechnology company working under a cloak of secrecy—ostensibly to protect proprietary information—might be tempted to develop new weapons for sale to the military, CIA, or another country or rebel army. The corporation, meanwhile, would risk no civil or criminal penalties by doing so. Whether or not such a scenario is plausible, the loophole contributes to international suspicion.

Domestic laws that prohibit such activities have been established by several nations, including the Soviet Union. In Great Britain the penalty for extreme violations can be life imprisonment and a fine of $10 million—enough to discourage any biotechnology corporation. Three bills have been introduced into the U.S. Congress since 1977. The most recent attempt was by Representative Peter Rodino in January 1987. So far all have foundered.

DANGEROUS ALLEGATIONS

The treaty's weaknesses are unquestionably significant. But allegations of treaty violations, when presented without convincing evidence, may be far more destabilizing. Since the late 1970s the United States has made numerous such claims against the Soviets. Many scientists and arms control experts believe the charges have consistently been apocryphal. This should not be confused with blind trust in Soviet activities or motives. The Soviet Union has done little to allay uncertainty about its commitment to the BW Convention.

The archetype for the U.S. accusations is yellow rain. As described in Chapter 3, despite years of U.S. efforts, few, if any, governments or international organizations support the assertion that the Soviets used TW. According to Meselson, even top U.S. officials were informed in 1986 by their own scientists that yellow rain "doesn't exist and never did." But the charges have been neither retracted nor softened.

Another case emerged from an incident in 1979 in Soviet city of Sverdlovsk, located in the Ural Mountains. For several weeks townspeople complaining of a choking cough and high fever flooded local hospitals. The unlucky victims—variously estimated at a few dozen to 1,000—quickly died of anthrax. When U.S. intelligence analysts learned of the outbreak in 1980, they immediately concluded that it had resulted from an explosion that released a cloud of anthrax spores from a secret BW factory.

The United States requested an explanation but proceeded to publicize the accusations the very next day—before the Soviets could reasonably have been expected to respond. The incident has often been recounted as evidence of Soviet treachery. After a wave of sensational news reports generated by the charge, the Soviets acknowledged the outbreak. But they blamed it on tainted meat, labeling the U.S. story a groundless fabrication.

Though rare in most industrialized countries, anthrax is endemic to large regions in the Soviet Union, including Sverdlovsk. The Soviets call the disease "Siberian ulcer." More than 150 outbreaks have occurred among animals in the Sverdlovsk province since 1936, and human outbreaks are expected in the Soviet Union about every three years.

Anthrax can be contracted through the pores of the skin (cutaneous anthrax, which is usually a minor problem), by eating contaminated food (gastric anthrax), or by inhaling spores (pulmonary anthrax). The gastric and pulmonary varieties are usually fatal if not treated immediately with antibiotics.

The U.S. charges, although based on circumstantial evidence, seemed far from spurious. The epidemic was unusually large, and the fatality rate extraordinarily high, according to some sources. A catastrophe of this magnitude is extremely rare in a nation with an advanced public health apparatus. Secondly, a Soviet émigré claimed to have received underground messages from dissident scientists confirming that Sverdlovsk housed a BW center. Perhaps most troubling was the fact that some victims suffered from systemic anthrax, including lung complications—considered a sure sign of pulmonary exposure.

But a number of scientists in the United States and abroad remained skeptical. Among these was one with unimpeachable credentials: Zhores Medvedev, a well-known dissident Soviet biochemist who emigrated to Great Britain. These critics made a strong circumstantial case against the charges.

"If an outbreak of pulmonary anthrax is the result of the accidental explosion . . . then the stories that the epidemic continued for a month, with thirty to forty casualties per day, could not have been true," said Medvedev. "Pulmonary anthrax develops a few hours after the infection has been inhaled, and the disease continues for only two to three days." If an explosion had occurred, Medvedev and others reasoned, authorities would certainly have rushed—for both security and humanitarian reasons—to administer antibiotics to the victims, a treatment that would have cut short the epidemic.

The U.S. case also overlooks the ability of gastric anthrax to poison the bloodstream, ultimately affecting all organs, including the lungs.

In October 1986, at the second review conference of the BW Convention, the Soviet representatives offered their most detailed explanation to date: Accidental distribution of contaminated grain infected privately owned cattle, which were subsequently slaughtered and sold on the black market. When the contamination was discovered, some "undisciplined workers" threw the animal carcasses into open garbage bins, thereby allowing some spores to be released into the air and necessitating decontamination by the army. The story is not farfetched, considering the scarcity of fresh meat in the Soviet Union. The existence of an illegal underground economy, combined with an apparent public health failure, is a major embarrassment for the Soviets. This may explain their initial reticence.

U.S. officials still imply that intelligence information supports their charges. But Meselson, who has reviewed classified data through his participation in a government study, remains unconvinced. He says every reliable indicator points to a natural outbreak.

In recent years on dozens of occasions high-ranking U.S. officials have accused the Soviets of operating up to seven secret BW research and production facilities at which genetic engineering is exploited in violation of the BW Convention. Every few months the DOD recycles these unsubstantiated charges. For example, in November 1986 a glossy, multicolor Defense Intelligence Agency booklet warned that the Soviets "have made remarkable progress in developing their biotechnological capabilities. Unfortunately, these same technologies are being used by the Ministry of Defense to develop new and more effective BW agents."

"We continue to obtain new evidence that the Soviet Union has maintained its offensive biological warfare program and that it is exploring genetic engineering to expand their [*sic*] program's scope," Secretary of Defense Caspar Weinberger said in justification of the Dugway aerosol lab. In making such statements, the DOD never presents any information that may be independently reviewed or analyzed. Convincing evidence of a Soviet rDNA threat is always classified.

And at no time has the United States lodged a formal complaint with the UN Security Council regarding these allegations, as directed by the BW treaty. Of course, the Soviet Union, as a permanent member of the Council, could veto any action it disagreed with. But failure to attempt even once to use existing legal remedies calls into question how seriously the United States takes the BW Convention and the veracity of American charges of Soviet infractions.

On some occasions, vague charges are leaked to friendly reporters, such as syndicated columnist Jack Anderson, who dutifully print the old claims as new and frightening developments. Sources for the stories—CIA or DOD intelligence operatives or bureaucrats—are never named.

The most dramatic and influential journalistic support for the U.S. case was made by *Wall Street Journal* editorial writer William Kucewicz in 1984. In a detailed series of eight articles entitled "Beyond 'Yellow Rain': The Threat of Soviet Genetic Engineering," he attempted to prove that "the Soviet Union is engaged in an intensive research program focused on using the revolutionary techniques of recombinant DNA to create a new generation of germ-warfare agents."

The series was notable for its deviation from certain fundamental journalistic standards. The articles were largely based on comments of unnamed intelligence sources or hearsay from émigrés who had not seen the inside of a Soviet lab in years. A primary source, Mark Popovsky, formerly a Soviet science writer, emigrated in 1977. Kucewicz concluded that a wide range of biomedical studies published in the open Soviet scientific literature is inherently offensive.

He mentioned only in passing, however, that research he considered inevitably oriented toward the production of novel BW agents in fact mirrored the U.S. effort in every respect. And the most "damning" statements attributed to Soviet leaders about the potential applications of geneteic engineering could have come straight out of the mouth of Caspar Weinberger.

When pressed, U.S. officials backpedal. They acknowledge that their secret data fall far short of proof. "It's not that we have affirmative evidence," Thomas Dashiell, one of the Pentagon's top CBW specialists, has said. "It's just that we don't know what's going on at certain laboratories."

A MATTER OF MOTIVES

This pattern of repeated, unsubstantiated, unverified allegations has troubled the world arms control community. In question are Reagan administration motives. Many analysts have suggested that the goal of such actions is not to bring the Soviets in line with international law but to generate domestic support for increasing CBW programs by portraying CBW treaties as impotent and the Soviets as international criminals.

"We need safeguards for allegations," not for treaty violations, according to Meselson. In part, this means conducting investigations of purported violations according to sound scientific procedures, such as competitive research teams—notably absent in the U.S. review of yellow rain.

Meselson has said CBW arms control would be well served by a U.S. admission that it was wrong about yellow rain. "People around the world would say, '*There* is one of the few governments on the planet that can admit error.' And our can," he added, "because we're not talking here about a value, or a principle—we don't believe in the principle of yellow rain—we believe in the principles of freedom, of honesty, of self-determination. . . . You demonstrate belief in those principles, and strengthen the principles by admitting error. It's in a totalitarian country that you can't admit error."

For their part, the Soviets occasionally make equally unverifiable countercharges. The United States is supplying the Afghan rebels with CW and shipping CBW to El Salvador, they say; the AIDS virus is an American BW experiment gone awry, according to a particulary mean-spirited claim by a New York-based Soviet journalist. Such accusations have not been pressed in a U.S.-style media blitz, but they certainly contribute to superpower distrust.

More important, the Soviet response to U.S. allegations has usually been defensive and condemnatory. The Soviets have often failed to assume a posture of open reassurance that they are, in fact, following the treaty in good faith. Their propensity for secrecy, together with this sullen reluctance to respond substantively to serious charges, has fueled speculation that the Soviets have something to hide. Increased openness under General Secretary Mikhail Gorbachev, however, is apparently affecting the BW sphere—notably the first full accounting of the Sverdlovsk incident, noted above.

SOME FORWARD MOTION

By the second BW Convention review conference in September 1986, little doubt remained that the treaty regime was unraveling. "Who can say, in the absence of compliance-verification procedures," asked SIPRI in a 1986 analysis, "whether the allegations regarding Sverdlovsk . . . or the yellow rain episodes were malicious propaganda, disturbing revelation or simply a consequence of heedless reliance upon unreliable intelligence?" SIPRI was so pessimistic about the prospect of making headway in a climate of hostility that it suggested that the review conference might be counterproductive.

These concerns gained credibility a month before the conference. A high-level U.S. negotiator, Douglas Feith, the deputy assistant secretary of defense for negotiations policy, condemned the treaty in the harshest terms yet. "The major arms-control implication of the new biotechnology is that the [BW Convention] must be recognized as critically deficient and unfixable," he told a congressional committee.

"Because new technology makes possible a massive and rapid breakout, the treaty represents an insignificant impediment at best," Feith continued. "It's principal failing, therefore, is no longer the absence of verification provisions or lack of effective compliance mechanisms, the commonly acknowledged shortcomings, but its inability to accomplish its purpose." Calling the treaty "a false advertisement to the world," Feith made comments that were nothing short of a signal that the United States was prepared effectively to abandon it.

Paradoxically, while the U.S. delegation to the conference reiterated its charges against the Soviets, it retreated from Feith's extreme position. The review session, in fact, achieved surprising progress in building confidence in the treaty process.

Perhaps the BW conference was the beneficiary of preparations for the 1986 Iceland superpower summit. "The superpowers appeared content to lie low at the conference so as not to let biological issues intrude on affairs they saw as more important," said U.S. molecular biologist Barbara Rosenberg, an observer at the conference. "The United States and the Soviet Union can usually afford to use biological questions as pawns in power politics, but under the circumstances they could not resist the constructive concern of other less powerful nations, who [*sic*] demonstrated that their concerted effort can carry the day in a multinational situation."

Whatever the reason, the conference proved to be the most encouraging development in biological arms control in many years. No actual changes were proposed for the treaty. Instead, a variety of measures was adopted to promote a political commitment to the arms control process.

The major accomplishment was a call for increased "transparency." The delegates mandated a broad sharing of information to "reduce the occurrence of ambiguities, doubts and suspicions." They agreed to establish procedures for exchanging detailed data on defensive BW research permitted by the treaty and on unusual disease outbreaks. The final declaration called for increased publication of research and urged "active promotion of contacts between scientists engaged in biological research directly related to the Convention, including exchanges for joint research on a mutually agreed basis."

In the spring of 1987 a thirty-nine-nation ad hoc committee of the review conference met in Geneva to codify how this information would be exchanged. If used faithfully, the detailed system it established will greatly ease suspicions on BW-related activities.

The conferees also strengthened and streamlined the provision on consultation. It will ostensibly be easier to convene a multinational meeting of experts to discuss possible violations without depending on the highly volatile UN Security Council. There was tacit agreement that the UN secretary-general be allowed to form such expert committees and dispatch them to trouble spots, as Javier Pérez de Cuéllar did in the Iran-Iraq case.

Reaffirming a 1980 conclusion, the delegates agreed that the convention covers all new scientific and technological developments. But they went a step further, explicitly defining *toxin* for the first time. "The Convention unequivocally applies to all natural or artifically created microbial or other biological agents or toxins whatever their origin or method of production," reads the final declaration. "Consequently, toxins (both proteinaceous and non-proteinaceous) of a microbial, animal or vegetable nature and their synthetically produced analogues are covered." This significant breakthrough appears to close the toxin loophole.

Finally, the declaration emphasizes the importance of each party's using its domestic laws to reinforce and validate the treaty's principles. These were impressive accomplishments in the face of superpower acrimony, achieved by seeking political commitments that enhance and clarify existing treaty obligations, rather than by pursuing legal changes. "Because formal amendments to the Convention are only binding on those states which accept them," said Falk, "negotiations to this end would likely produce a dual-track [multitrack, in the case of more than one amendment] treaty regime with different states having quite distinct obligations."

Before the second review conference, experts feared that opening the treaty to formal amendments could backfire, weakening restrictions or raising questions that defy resolution. In biological research, comprehensive verification, for example, would be incredibly intrusive and burdensome. "You'd have to inspect the lab notebooks of every lab in the country," according to Richard Novick.

"If nominal verification machinery was agreed upon, then it could be easily evaded," Falk noted. "As a consequence, attainable levels of verification are likely only to shift the confidence problem, and in certain circumstances might even aggravate it." The second conference proceeded so smoothly that formal amendments began to look more appealing. In the final analysis, however, the delegates feared that legal changes would bog down the process, forestalling the all-important information exchange.

The measures adopted were voluntary. But they have the power of unanimous acceptance, guaranteed by each party's need to retain political credibility. In any case, the treaty depends on good faith. As mandated by the second conference, by 1991 the third review conference will be convened and must evaluate the effectiveness of political measures as well as reconsider legal changes.

Without these gains being minimized many problems remain. Biochemical weapons, such as fast-acting hormones, and prototype research are apparently permitted for purposes of gaining defensive insights. Some means of restricting these prospects are vital. Where can lines be drawn between offense and defense in research and development? Should BW research be internationalized to ensure full disclosure and defensive integrity? The conference avoided each of these essential questions for good reason: Mutually acceptable answers may not be forthcoming in the foreseeable future.

Part of the problem is a failure to establish treaty institutions. "At some stage," wrote Nicholas Sims, who represented Great Britain at the first review conference, the parties "must come to appreciate the need for some kind of intergovernmental committee to oversee the working of the Convention, on a permanent basis. . . . A healthy treaty regime cannot be had on the cheap."

THE FUTURE OF INTERNATIONAL CONTROLS

The long-term vigor of the BW disarmament regime rests largely on superpower intentions. Contradictory statements and actions have made these intentions enigmatic. The Soviet stance on verification is a case in point. After consistently resisting specific verification measures, the Soviet delegation at the 1986 review conference, apparently buoyed by Gorbachev's influence, reversed its position by proposing negotiations for a supplementary protocol to create strong verification procedures.

"Was the Soviet Union suddenly . . . calling the bluff of those who had been loudest in deploring the [treaty's] paucity of verification provisions . . . and its consequent vulnerability to violation with impunity. . . ?" asked Sims. Or was the proposal serious and legitimate? In either case, he concluded, it "signalled a marked reluctance on the part of the Soviet Union and its allies to allow further [legal] obligations to be tacked onto the Convention without fresh negotiations."

This demonstrates that the Soviet position is changeable, even unpredictable. But the Soviet government speaks with one voice. In this regard the United States is more problematic. U.S. officials publicly agree that the treaty should be respected, but they appear to be at odds about its scope and significance in biological arms control.

Some American arms control leaders have been in the forefront of promoting and strengthening the treaty regime. Arms Control and Disarmament Agency (ACDA) official Robert Mikulak, a member of Pugwash, has consistently called for greater openness. "There is no justification for classified military research on recombinant DNA—in the United States, the Soviet Union, or anywhere else," he wrote in 1984, echoing other U.S. statements.

At the same time, however, when they are not busy pressing unsubstantiated assertions of Soviet unreliability (and even Mikulak has joined this bandwagon), many U.S. policymakers engage in "treaty bashing." Feith's comments are by no means unique. In 1985 the Defense Science Board on Chemical Warfare and Biological Defense concluded that "technology has made obsolete much of the distinctions and language of the BW treaty." The ink was barely dry on the review conference declaration before H. Allen Holmes, an assistant secretary of state for politico-military affairs, was quoted as saying, "The Convention, in our judgment, cannot be made effective through amendment or design."

The U.S. position on creating prototype BW for defensive inquiry, as discussed in Chapter 5, is similarly ambiguous. In 1969 a DOD spokesman told a congressional committee, "Without the sure scientific knowledge that such a weapon is possible, and an understanding of the ways it could be done, there is little that can be done to devise defensive measures." This statement clearly reflects the view that novel agents are essential to BW defense.

In recent years ACDA and U.S. Army officials have publicly ruled out the development of prototype weapons systems. But other statements (such as Weinberger's assertion of the need to test "known and suspected agents" to counter the Soviet biotechnology threat—see Chapter 6) strongly suggest that the 1969 view represents today's actual policy. And in a report about Dugway to a congressional committee, the army stated: "The result [of biotechnology] is a manyfold increase in the number of candidate agents. . . . The potential agents must be analyzed and characterized. . . . This itself is a formidable task and requires working with the hazardous materials."

The implication that prototype weapons are being developed or planned is inescapable. What are enemies of the United States to believe? If such statements were made by the Soviets, there would be no confusion. The United States would immediately charge that the Soviets intended to use such weapons.

THE QUESTION OF LINKAGE

Other problems stand in the way of biological arms control. "There is no doubt that, as [biotechnology] matures, so will its application to CBW weaponry become increasingly practicable," SIPRI noted in 1985. While this is taken as justification for large-scale defensive research, the offensive-defensive tautology will inevitably place new pressures on the treaty process.

In the long run better superpower relations and progress on overall arms control must form the basis for stability and increasing controls in the BW realm. Because of the similarities between biological and chemical warfare theory, signing a CW convention is of particular importance. Negotiations for such a ban have plodded along for years, although, as noted in Chapter 4, there have recently been some hopeful signs. The draft CW convention incorporates highly advanced consultation and verification measures and could serve as a model for improvements in the BW treaty.

It bears repeating that Soviet secrecy has been a major impediment to progress on chemical and biological arms control. But hopes for a CW treaty may actually be threatened more by Washington. The Reagan administration has suggested that any new treaties with the Soviet Union are seriously hampered by the yellow rain and Sverdlovsk affairs, which at this point are static issues for the United States, not open for reconsideration. The larger question is whether any Reagan administration arms control initiative can be taken seriously.

Strident U.S. accusations of Soviet treaty violations have by no means been limited to BW. On many other agreements, including the SALT II and antiballistic missile (ABM) accords, a steady drumbeat of similarly questionable charges has been sounded. In each case the loudest protests against such claims have come not from the Soviets but from U.S. arms control experts, legislators, and allies who challenge the administration's stance. In a number of cases departments within the administration itself have rejected White House assertions.

The significance of these actions and attitudes to the CBW arms control process is analyzed in detail in the concluding chapter. But a change of governments—even to one that is less anti-Soviet and more compromising—will by no means guarantee a more constructive approach to CBW. "Most weapons have presumably been considered unconventional in their time, and moral outcry has generally attended the introduction of novel ones. Moreover, the military has often been reluctant to assimilate radically new techniques of fighting," SIPRI commented in its landmark 1972 study of CBW. "But there is an equally long tradition of military expediency overcoming moral resistance or conservative reaction to new methods." American attacks with CS gas and herbicides in Vietnam and the Iraqi use of mustard gas validate this analysis.

Incited by a combination of faulty intelligence and its own propaganda, a future government of the United States or the Soviet Union might be swayed by Feith's view that the BW treaty represents "a false advertisement." The logical extension of this argument is that participation in ongoing BW negotiations is in itself a dangerous charade. Furthermore, a future superpower government might feel compelled to withdraw from the treaty, effectively killing it. This extreme step might be motivated by genuine conviction or intentional sabotage. Regardless, such a move would be justified as essential to bolster the integrity of true arms control in an often lawless world.

The BW Convention is far from perfect. But it does play an important role and shows signs of being adaptable to modern challenges. The pact needs and deserves the aggressive support of an informed public, particularly in the face of an often reckless and apparently cynical U.S. administration.

8

REGULATING NEW LIFE-FORMS

A CAUTIOUS BEGINNING

International law presents at best a porous barrier against CBW research that may be dangerous or destabilizing or illegal. What about domestic regulatory controls?

Biotechnology regulation began with a yearlong self-imposed moratorium against certain potentially dangerous rDNA experiments, initiated by leading scientists at a historic 1975 meeting at the Asilomar Conference Center. The culmination came the following year, when the National Institutes of Health issued "Guidelines for Research Involving Recombinant DNA Molecules," based largely on the Asilomar talks. The guidelines constitute the basis of all rDNA lab regulation in the United States.

In formulating the guidelines, the NIH was forced to address the unique character of biological organisms. That is, a toxic chemical, once released into the environment, has finite effects. Chemical spills almost always degrade or can be cleaned up, although this may be difficult and time-consuming. *Biological* organisms are different. Given nurturing conditions, a single bacterium can double every twenty minutes. In as little as a day it can give rise to up to 5 sextillion (5 followed by 21 zeros) progeny bacteria. They cannot be recalled, and under plausible scenarios human beings would be hard pressed to stop the relentless proliferation.

Correspondingly, the initial guidelines were extremely stringent, at least in retrospect. They banned six classes of work, including rDNA manipulations involving many cancer viruses, pathogens, and genes coding for the expression of lethal toxins. The environmental release of recombinant organisms and their production on a large scale were also prohibited. And the guidelines strictly limited transfer of recombinant plasmids to a single bacterial host—a strain of *E. coli.*

The logic of the NIH guidelines is based on dual containment. Physical containment is defined by four levels, designated P1 through P4, corresponding to the relative hazards of the organism being altered. P1 amounts to little more than using standard lab practice and common sense, while researchers in superinsulated, negative air pressure P4 labs wear what could be the garb of astronauts. They scrupulously follow elaborate, redundant safety procedures to keep themselves free of infection and prevent recombinant organisms from escaping into the environment.

Biological containment relies on using "fail-safe" bacteria that would reliably perish outside artificial lab conditions. When the guidelines were established in 1976, this combination of restrictions and safety methods persuaded most scientists that an ecological or public health disaster caused by recombinant DNA was extremely improbable.

The guidelines also mandated institutional biosafety committees (IBCs), registered with the NIH, for all research entities—including universities, government agencies, or corporations—conducting rDNA experiments. The local committees were designed to ensure that all work was in accordance with the guidelines. Finally, the NIH created the Recombinant DNA Advisory Committee (RAC). The RAC is comprised of molecular biologists and other scientists as well as specialists in law, public policy, and environmental affairs. While the director of the NIH makes final decisions about the guidelines and their interpretation, the RAC effectively functions as the Supreme Court of rDNA research.

The term *guidelines* was chosen with care. It was emblematic of this historic effort by scientists to develop effective self-regulation. The guidelines are, in fact, mandatory for projects funded by some agencies, including the NIH, while other federal agencies comply voluntarily. The NIH compels the compliance of grantees by threatening to withdraw offenders' federal support. Privately funded biotechnology companies—virtually nonexistent in 1975 but now numbering in the hundreds—have never been legally bound by the guidelines. It is believed that the vast majority have accepted the standards without incident.

ADAPTING TO REALITY

The scientific community soon realized that many initial fears about the dangers of recombinant organisms were exaggerated. Almost from the day the NIH guidelines were issued, the agency felt tremendous pressure for reform. The guidelines were described as overzealous constraints on progress. Under the original provisions the incipient biotechnology industry would never have gotten off the ground and academic research would have been crippled. The guidelines had been conceived, however, on the assumption that after experience with the technology and careful deliberations, changes would be adopted as warranted.

From 1976 to 1982 the guidelines underwent a series of sweeping revisions. All the original prohibitions have been eliminated, while the range of organisms allowed as hosts or donors is now virtually unlimited.

Although the categories of physical containment are still on the books, actual requirements for their use have been vastly reduced. Up to 90 percent of all rDNA experiments can now be conducted at the P1 level, leaving only the most hazardous pathogen and toxin experiments at P2 to P4. Only four facilities in the nation now maintain P4 labs, the primary one being operated by the army's BW researchers at Fort Detrick.

Most experiments that formerly required RAC approval now need only IBC approval. Studies once subject to IBC assent can proceed after the researchers merely notify the committee, while 80 to 90 percent of rDNA work is completely exempt from review. A proposal to make the guidelines voluntary even for NIH-funded scientists was narrowly defeated in the RAC. (It's important to note that only biology's ultimate exponent, rDNA, has been subjected to a full-blown regulatory process. Other powerful new technologies, such as monoclonal antibodies, have been considered benign and have never been specially scrutinized.)

Clearly many of the changes were essential for scientific and commercial development. One signal of this is that several other nations have largely followed the NIH lead in both the development and the relaxation of their own rDNA rules. The majority of researchers believed the revisions struck the proper balance between respect for the unknown and the reality that most rDNA applications are totally safe. The original, more stringent guidelines were founded on concerns over potential hazards that simply did not materialize after years of cautious experience.

But a number of prominent scientists believed the changes went too far—both scientifically and as an instrument of social policy. Robert L. Sinsheimer, former chancellor of the University of California at Santa Cruz and a biophysicist, played a key role in developing the guidelines in 1975. "I think they've been diluted to a point where they're almost meaningless," he said.

In a letter to the NIH protesting the proposals that led to the major revisions, Sinsheimer wrote: "The net result of [the changes] is to dismiss the possibility of hazard from all recombinant DNA experiments except those involving *known* [emphasis in original], very pathogenic agents . . . to dismiss the possibility of the creation of a novel pathogen, or of a more virulent form or a new mode of dissemination of an existing pathogen."

MIT's Jonathan King told the NIH that the changes send the wrong signals to the emerging biotechnology industry: "Outside of a very few proscribed experiments, any of a vast variety of modifications of organisms can be carried out without forethought, without special precautions, and without informing any agency of government of their nature; that there will be no sanctions if in fact irresponsible and dangerous procedures are engaged in; and that no long-term epidemiological, monitoring, risk-assessment program need be put in place in parallel with the developing new technology. That is, anything goes."

Not surprisingly, the RAC rejected these concerns as overcautious or radical. While any government regulation of a technological process seeks to protect public safety, reasonably rapid economic and scientific advances are also vital societal goals that deserve protection. Considering the overwhelming evidence of responsible behavior by the scientific community, the RAC reasoned, relaxing the guidelines was in society's best interest.

For its part, any emerging industry tends to portray initial critics as Luddites, sensationalists, or antibusiness alarmists, as consumer advocate Ralph Nader has pointed out. While industry attempts to appeal to society's hopes, critics appeal to fears—a much harder way to win hearts and minds.

The RAC also saw the handwriting on the wall: Restrictions widely regarded as unreasonable were bound to be flouted. Even relatively mild limits were transgressed without significant repercussions—by the overeager, the careless, or scientists convinced of the safety of their work and its overriding importance to humankind.

In 1980 researchers in the lab of University of California at San Diego Professor Samuel Ian Kennedy accidentally cloned genes from a prohibited pathogenic virus. During an NIH investigation, which uncovered other experiments that had not been approved by his IBC, Kennedy resigned his post. Although the violations were unequivocal, the NIH could do little else to punish Kennedy, who was no longer employed by the university.

In the same year Martin Cline, professor of hemotology and oncology at UCLA, proposed human gene therapy experiments designed to treat sickle-cell anemia and beta thalassemia—diseases that, in certain forms, are fatal and incurable. Both are caused by dysfunctional hemoglobin genes. Cline, who planned to treat his patients with corrective genes, gained approval after a long, exasperating regulatory battle. But at his moment of victory Cline changed the experiment. Instead of inserting the naked hemoglobin genes into his patients, he inserted them into a recombinant plasmid without obtaining approval from his IBC. When this violation became known, Cline was stripped of one NIH grant. However, two other funding agencies did nothing. And he remained a full professor.

For their part, legislators have largely steered clear of rDNA. They have favored the flexible, evolutionary approach of the NIH over the rigidities that could be imposed by new legislation. And the preservation of freedom of inquiry and economic development, they surmised, was better served by leaving things up to the scientific experts.

By 1983 the regulatory controversy had moved out of the lab and into the farms and hospitals. At this point the biotechnology business had many billions of dollars invested in the development of scores of new pharmaceuticals and agricultural products. In order to market these products, they had to be field-tested.

The regulatory process soon grew more complicated and congested. New federal referees, such as the Environmental Protection Agency and Department of Agriculture, began to assert their regulatory authority over biotechnology. The prospect of deliberate release of recombinant organisms into the environment renewed the most profound question related to lab accidents: What is appropriate protection against a low-probability but high-risk event—rDNA environmental catastrophe?

FROM ICE-NINE TO ICE-MINUS

During 1985 and 1986 recombinant agricultural products were tested in gross violation of federal rules on deliberate environmental release of rDNA. The most graphic example involved recombinant strains of the common bacteria *Pseudomonas syringae* and *Pseudomonas florescens.*

The new organisms were developed by a University of California scientist using gene deletion—removal rather than addition of genes. Advanced Genetic Sciences (AGS), a biotechnology company based in Oakland, California, took out a license on the new organism for possible marketing under the name Frostban. Dubbed "ice-minus" in the press, the bacteria colonize on leaves, inhibiting frost formation when temperatures drop to a few degrees below freezing. Crops now destroyed by minor cold snaps could theoretically survive if sprayed with ice-minus.

Initially the EPA issued a field test permit but then rescinded approval after revelations that AGS had violated the agency's rules by injecting ice-minus into trees on the roof of its corporate headquarters. The agency fined the company $13,000—the first EPA penalty levied for gene-splicing tests and one of the largest fines ever ordered for a violation in agricultural research.

Fear of molecular alteration is deeply rooted in the public mind. In *Cat's Cradle,* a popular 1974 novel by Kurt Vonnegut, Jr., scientists develop a form of water that freezes at room temperature. "Ice-nine," which allows troops to cross rivers on solid ice, even in summer, eventually destroys the earth.

Nearly every informed person in this debate views as farfetched the prospect of a wildly proliferating microbe laying the ecosystem to waste. AGS and many academic scientists complain that some critics nevertheless fuel controversy by playing on such public fears. They argue that genetic manipulations are akin to age-old crossbreeding, naturally occurring mutation and gene recombination, or the use of live attenuated viruses as vaccines.

But neither science fiction nor genetic engineering need be invoked to counter these points. For example, Dutch elm disease, which virtually eliminated the American elm from many parts of the country, and the blight that similarly destroyed the American chestnut tree both were caused by the unintentional introduction of existing organisms into new ecological niches.

Admittedly these were unusual cases. Even the most strident critics of the release of recombinant organisms agree that the vast majority of altered microbes are unlikely to produce significant, unwanted secondary effects. But even the remote possibility of a recombinant organism's causing problems analogous to the Dutch elm episode, they argue, mandates extreme caution for now.

Most scientists doubt that microbes released in a small test could survive long in the environment. Genetically altered organisms usually are unlikely to compete successfully against infinitely more numerous, naturally adapted microbes. But no scientist will categorically rule out the possibility of permanent displacement, particularly in agricultural applications over thousands of acres.

When the AGS controversy became front-page news in the fall of 1985, Eugene Odum, a University of Georgia ecologist, wrote in a letter to *Science* magazine that *P. syringae,* like a few other natural bacteria, enhances rainfall. Odum cited a study that indicates the microbe is "wafted up into the clouds" where it forms the nuclei for raindrops. This is the precise property genetically excised from the ice-minus strain of *P. syringae* to prevent frost damage.

If wild *P. syringae* "does indeed have a beneficial role in enhancing rainfall," Odum wrote, "then the ecologist's concern about possible secondary or indirect effects of releases of genetically altered organisms (such as reduced rainfall after massive use of ice-minus) is vindicated—incredibly, at the very first major controversy over release of engineered organisms."

Still, many noted ecologists, such as UC Berkeley's Robert Colwell, a member of the EPA's ice-minus evaluation group, believe that the organism is safe for field testing. But Odum's point is important—perhaps not so much for this single case as because ice-minus is setting a historic precedent. In a few years hundreds of recombinant organisms will be ready for testing, if not for full-scale use. The creation of a regulatory system that efficiently can handle this volume of products, without losing the public's trust, is therefore vital, critics of field testing say, to be reasonably sure that not even one in a thousand altered organisms will cause ecological or public health problems.

Most ecologists agree, however, that the science of predictive ecology, on which such a regulatory process must rely, is in its infancy. There is no consensus about how much and what types of data are required to make decisions about the effects of altered organisms. Until this science is evolved, the release of novel life-forms will always entail a level of risk considered unacceptable by some sectors of society.

Ironically, the insurance industry represents the strongest prospect that such concerns will ultimately be taken seriously by regulators and legislators. Hardly any biotechnology firm or research institution in the United States has insurance for field tests involving rDNA, according to Jeremy Rifkin, an outspoken critic of genetic engineering. Without such coverage, the potential financial damages caused by an ecological debacle, however remote the prospect, would be prohibitive.

When the nuclear power industry found itself in an analogous position in the late 1950s, it lobbied Congress successfully for the Price-Anderson Act, which explicitly limits industry liability. Rifkin filed suit in 1986 to compel the EPA to enforce laws that require adequate insurance coverage and to halt field testing until the question is resolved.

Amid these discussions, AGS, a multinational corporation, threatened to take its field tests—and its business—abroad if the EPA didn't back off. "What matters is that the research go forward," Joseph Bouckaert, the chief executive of AGS, told the *Wall Street Journal*. "We will move forward, if not in this country, then in Europe or Latin America."

His statement was a vivid reminder that the compelling forces of corporate survival—rather than overaggressive academic researchers—place the greatest pressure on methodical, careful regulatory development that speaks to society's fears as well as its hopes.

A few months later, in the summer of 1986, Bouckaert's threat was carried out—not by AGS but by the nation's oldest biomedical research institution, the Wistar Institute of Philadelphia. Wistar tested a live, genetically altered rabies vaccine on cows in Argentina without the consent or knowledge of the Argentine or U.S. governments. When the test was revealed months later, the Argentines were outraged. According to a report in the *New York Times*, "Farm workers who handled the animals and milked them had never been told that the herd had been inoculated with an experimental virus."

Two days after this revelation an Oregon State University research group went public about its own recombinant animal vaccine tests, conducted in New Zealand in April 1986.

A NEW BEGINNING

Long before these incidents had run their course, the confusing regulatory turf battles being waged by various federal agencies prompted industry representatives to start complaining bitterly about the difficulties of bringing products to the public. The Reagan administration turned a sympathetic ear, and in June 1986 the president signed a directive establishing a new framework for biotechnology regulation, ostensibly designed to clarify vague, overlapping jurisdictions.

David Kingsbury, the former director of a major U.S. Navy BW research laboratory and now an assistant director of the National Science Foundation and architect of the new policy, claimed it results in more effective oversight than before. But there is cause for skepticism. When Kingsbury heard about the Argentina incident, his response was: "Wistar must have felt that the regulatory structure was too stringent. We may be overregulating and pushing companies to test their products overseas."

Although it defines jurisdictions, the new framework ignores regulations whose ambiguous language makes it easier to dodge the law. AGS claimed, for example, that injecting ice-minus under the barks of trees did not constitute an environmental release as defined by the EPA. And according to a lengthy analysis released by the congressional General Accounting Office in 1986, Department of Agriculture rules are replete with glaring flaws and gaps. The GAO recommended major clarifications and revisions.

Most significantly, the new framework assumes deletion products, such as ice-minus, and alterations involving "regulator genes," which control the functions of other genes, are benign. Both categories are virtually exempt from field test review.

In general, industry and many scientists applauded the new framework as a progress-oriented approach that still safeguards the public. But other scientists were shocked at the oversight exclusions. Regulator genes, for example, perform vital modulation of the production of proteins by a cell. The new rules treat such genes as if they were "biologically unimportant," said Liebe Cavalieri, a molecular biologist and cancer researcher at the Sloan-Kettering Institute in New York. "The administration's position is scientifically indefensible."

Similarly, many scientists believe the deletion of a single gene may upset the delicate balance of relationships between genes, leading to mutations, wild multiplication, or other unforeseen consequences. "It is a medieval scientific view that a deletion is automatically less risky," said MIT's King. "This decision has the look of a response to commercial pressures to weaken regulations, rather than seriously making an effort to maintain protections."

Such concerns are particularly relevant in light of a move by the White House Council on Environmental Quality earlier in 1986. The council scrapped the rule that required federal agencies to consider the most extreme environmental impacts—the "worst-case scenario"—of their decisions, on the basis that such consideration unnecessarily breeds "endless hypothesis and speculation."

In the absence of broad data with which to draw firm conclusions, environmentalists once used this rule in attempts to stymie what they perceived as dangerous deregulation. It is also the reasoning that critics of deliberate release of rDNA were compelled to rely on. But now the Reagan administration seems to be trying to pretend that worst-case scenarios do not exist.

MILITARY VOLUNTARISM

Where does the Department of Defense fit into this schema? Generally the DOD is exempt from any regulation promulgated by another government department or agency. But regarding the NIH guidelines, the DOD volunteered to comply and specifically ordered its branches and contractors to do so. It has become increasingly clear over the past few years, however, that the DOD's voluntary compliance, even within the generous parameters of the NIH guidelines, has been inconsistent at best.

A key provision of the guidelines is maintenance of an IBC by every institution conducting rDNA studies. In view of the weakened role of the RAC over the years, these bodies have become the sole checkpoint for rDNA work involving a wide range of pathogens. Our spot check of the DOD's list of approved rDNA studies and descriptions of other experiments, obtained under the Freedom of Information Act, showed a cavalier disregard for this rule, however. Three universities—Brigham

Young, Tulane, and North Carolina—and three corporations—Martin Marietta Laboratories, Syntro, and MicroGeneSys—that conducted rDNA work for the military in 1985 and 1986 failed to register IBCs with the NIH as required. Their studies involved everything from adhesives to anthrax.

Two institutions—Tulane and North Carolina—lost NIH approval of their IBCs in 1985 for ignoring requests for information on their committees and rDNA activities, according to NIH official Stanley Barban. The others never bothered to file applications for recognition of a safety committee, he added. In 1982 the Illinois Institute of Technology conducted work for the DOD related to nerve gas detoxification before it registered an IBC. It finally did so in October 1983.

Determining if these committees are properly constituted and registered is one of the few noninvasive ways to check DOD adherence to the NIH guidelines. By repeatedly ignoring this fundamental safety requirement, the DOD casts doubts on its overall commitment to accepted lab practices and limitations.

FACING THE REGULATORS

On a handful of occasions the RAC has been forced to address the biological weapons implications of work with rDNA. In each case its actions have been ineffectual—and by some standards counterproductive.

In June 1982 a heated debate broke out within the RAC. Committee members Richard Goldstein of Harvard and Richard Novick of the New York Public Health Research Institute sought an amendment to the NIH guidelines prohibiting "the construction of biological weapons by molecular cloning." They were motivated by concern that a permissive interpretation by the DOD of the 1972 BW Convention could lead to nefarious use of rDNA.

"An explicit public statement [could] have several important effects," the scientists wrote. "It will provide automatic public support for a refusal of the scientific community to participate in the development of biological weapons" as well as bolster national and international confidence in the treaty regime.

The proposal was meaningful, in part, because of the absence of domestic legislation prohibiting BW development by private citizens, corporations, or universities, although such legislation is required under the terms of the 1972 BW Convention. The scientists also hoped it would reduce the potential for the military to define all manner of laboratory work—even the construction of recombinant prototype agents for threat assessment—as defensive and therefore permitted by the treaty.

The RAC, persuaded by reassurances from the army and the ACDA that the 1972 treaty effectively bans recombinant BW, rejected the Goldstein-Novick proposal. In doing so, the committee overlooked both the lack of domestic implementing legislation and another key concern: Even if today's military leaders are completely trustworthy and aboveboard, future policymakers might be tempted to adopt a more liberal interpretation of the treaty that would justify a far broader range of rDNA activities. This possibility becomes more likely as ongoing scientific advances further entice military planners already eyeing biotechnology with interest.

In October 1982 the RAC approved a study by Harvard University researchers working in P4 labs at Fort Detrick to clone the gene for diphtheria toxin—among the world's deadliest poisons. The purpose of the experiment, said the researchers, was to create a new therapy to treat melanoma, a type of skin cancer.

The committee dealt almost exclusively with the question of whether such an experiment could be conducted safely. It ignored the toxin's potential use as a weapon, the dangerous precedent of constructing a novel pathogen, and how the experiment might be perceived by other signatories to the BW treaty. Also, approval was given in the face of evidence presented by some scientists that the desired information about treating melanoma could be obtained in safer, less controversial ways.

In February 1983 the RAC received a DOD proposal to clone the gene for Shiga toxin, the causative agent for a common form of dysentery. The purpose of the experiment, said the researchers, was to create a vaccine. Paul Warnke, ACDA director during the Carter administration, formally objected. He pointed out that like the diphtheria example, this experiment could create a biological weapon.

Warnke was joined by genetic engineering critic Rifkin; Herbert Scoville, a former ACDA official as well as a former deputy director of the CIA; Richard Falk, professor of international law at Princeton University; and others in asking the RAC to table the Shiga proposal until after the DOD had provided an "arms-control impact statement." Such a statement is required under the Arms Control and Disarmament Act for any "program involving technology with potential military application."

The RAC overrode these concerns and approved the DOD experiment.

In May 1985 Rifkin tendered another proposal: that the RAC establish a subgroup to study the potential BW applications of rDNA technology, whose objective would be to advise the government and the general public on the issue. The committee declined to consider the idea on the ground that such a group would exceed its resources and scope.

During the controversy over the proposed Dugway Proving Ground P4 lab for research with large quantities of aerosolized pathogenic organisms and toxins (described in Chapter 6), the army never ruled out rDNA work as one of the lab's goals. In fact, responding to putative Soviet work in genetic engineering was the stated purpose for the lab. The army's own reports even indicated that rDNA would be an almost inescapable extrapolation of the planned research program. Such work, of course, would be antithetical to the NIH guidelines, which mandate absolute avoidance of pathogenic aerosols because of increased risk that some organisms would escape. But the RAC was never consulted on this prospect.

In none of these cases did the RAC—the premier scientific arbiter on rDNA—assume a leadership role. Part of the problem was a failure by the NIH to encourage the committee to consider the arms control implications of rDNA decisions. The agency has never provided the RAC with the DOD's annual reports on CBW activities—the only broad public accounting of military rDNA research. "How can they make an intelligent decision," asked Goldstein in an interview, "if they don't know what's going on?"

To King, however, this is an inevitable consequence of RAC's design. "I don't think you can ask a committee that's constituted on narrow technical grounds around safety to deal with questions of policy," he said. "We don't have an oversight committee for military genetic engineering, and I think we need one. The Arms Control and Disarmament Agency could do it, but they are [*sic*] inactive. . . . The RAC is trying to help the technology along. They [*sic*] don't want questions raised in public."

Some environmental laws do bind the DOD. They have been used effectively to postpone some advances in the BW research program. In November 1984 (as noted in Chapter 6) Rifkin and retired Rear Admiral Gene R. La Rocque, director of the Center for Defense Information, sued to stop the Dugway aerosol lab construction on the ground that the army had not complied with the provisions of the National Environmental Policy Act (NEPA). They were supported by Scott Matheson, governor of Utah.

NEPA requires preliminary assessments and in some cases detailed environmental impact statements (EISs) for many construction projects and research programs. The lawsuit pointed out that the army had not assessed the possibility that BW agents could leak from the proposed lab. It also charged that the army violated the federal Administrative Procedure Act by failing to consider adequately reasonable alternatives to using actual BW agents—for example, the use of biological simulants.

In response, the army quickly offered a cursory environmental assessment, along with a rationale explaining its need to test actual BW agents. But U.S. District Judge Joyce Hens Green rejected the assessment as inadequate. She wrote:

> Merely citing the safety features of a proposed facility without carefully analyzing the possible environmental dangers associated with the [environmental assessment] does not constitute the type of environmentally informed decisionmaking that the drafters of NEPA had in mind. The Environmental Assessment represents but an amalgam of conclusory statements and unsupported assertions. . . .
>
> The possibility of an accident involving personnel, or exposure to the outside environment, while low in probability, does exist. . . . Such an accident could produce extraordinary, potentially irreparable consequences.

Green then granted an injunction against the facility pending a full EIS. The decision has so far set the army back more than two years. More significantly, it showed the BW program to be vulnerable on legal grounds linked more to public health than arms control.

Rifkin used the same strategy in 1986 in a more ambitious effort to derail the entire BW defense research program. This suit called for an EIS covering all of the DOD's work involving BW agents as well as for new regulations for handling and keeping track of dangerous microbes. In a remarkable victory for Rifkin, in February 1987, the DOD agreed to prepare the sweeping review of its program. The decision represents one of the broadest applications of NEPA in its history. The EIS will take two years or longer to prepare.

Important as these legal victories are in the short run, they are likely to be ephemeral. Most informed observers believe that if necessary, the DOD will make the minor program changes and file the appropriate reports needed to prevail. The ultimate battleground for these issues will be not a courtroom but the offices and halls of policymakers. A case in point is a November 1969 law on open-air testing of lethal chemicals and BW agents. Such testing is essential to an advanced offensive program.

Congress forbade open-air tests if the secretary of the Department of Health and Human Services determined they could not be done safely. But the president may disregard the DHHS veto in light of "overriding considerations of national security." Although the president is required to report and explain such a determination to congressional leaders "as far in advance as is practicable," no specific time frame is attached to the obligation. In this way secret testing is indirectly permitted.

A week after the statute was passed, President Nixon renounced all use and possession of biological weapons. Therefore, the law paradoxically allows the president to approve open-air testing of weapons that are banned by presidential order.

CORPORATE CULTURE AND THE DOD

As a rule, regulations and regulators are profoundly influenced by the needs of industry. As previously mentioned, this is neither accidental nor even particularly controversial. It is by design. The biotechnology industry is no exception.

Many critics have suggested that corporations exert undue influence over regulators. The classic case is the petrochemical industry's bullying of government environmental agencies, resulting in devastating pollution problems. The concern also applies to biotechnology. While the RAC has unquestionably been stocked from its inception with scientists of unimpeachable technical qualifications, key players have also had a compelling financial stake in the outcome of the committee's deliberations.

The most obvious example, among many, is David Baltimore, professor of biology at MIT. Baltimore's scientific credentials, including a Nobel Prize for his contributions to rDNA research, arguably made him the most influential member of the RAC when it recommended dramatic reforms of the NIH guidelines. The most outspoken advocate of the changes, he sponsored the nearly successful proposal to make the guidelines completely voluntary.

"One thing that bothers me particularly is that David Baltimore should still be in position to advise government on these matters without disclosing his new deep conflict of interest," Harvard Emeritus Professor of Biology George Wald—also a Nobel laureate—wrote to the NIH during the debate over guidelines reform in January 1982. "He is the second largest personal stockholder in Collaborative Genetics, which was to go public this month, with holdings valued at over $5 million."

Economic conflicts of interest extend well beyond the RAC. "Most of the leading molecular geneticists, on whom we must depend for our understanding of the products generated by genetic engineering, have commercial affiliations," pointed out Sheldon Krimsky, a former member of the RAC and a leading historian of rDNA at Tufts University. "This may be a factor in how they assess the potential risks of genetically engineered life forms."

And regulatory architect David Kingsbury came under criminal investigation by the U.S. Justice Department in the fall of 1987 for allegedly hiding his ties to the biotechnology industry.

In addition to this direct financial interest in biotechnology firms, industry funds may support up to 25 percent of all biotechnology research in American universities, according to two 1986 studies published in *Science* magazine. Among faculty that receive industry support, 28 percent get at least half their research funds from corporate sources.

The attitude a regulator such as the NIH maintains toward the DOD is much the same as its perspective toward, say, Collaborative Genetics. In each case the NIH seeks to safeguard the public—but narrowly and in a manner that expedites scientific and industrial development. The DOD and industry alike benefit from the permissive climate such a regulatory process generates.

"The biotechnology industry and the genetic engineering community as a whole could play a vital role in raising public awareness of the dangers of military applications of biotechnology," King has said. "Instead, they deny the problem exists except in the minds of a few misguided or naive individuals. They are doing everything they can to discount BW and environmental hazards. Rather than taking the lead in exposing potential dangers, they are obscuring the issues."

Ironically, while corporations engage in life-or-death competition in product development, they collude with each other when their actions transgress regulatory boundaries. Competing companies have not moved to isolate even obvious offenders, such as AGS, because they want to downplay the significance of actions they emulate. Some corporate officials, of course, endorse a soft line of criticism, saying that these companies showed poor judgment and perhaps deserve a reprimand. But the vast majority either is silent or blames the regulations as unnecessary, even provocative. Furthermore, the regulators themselves, such as White House adviser Kingsbury, are often won over by these arguments.

The DOD does not function in complete isolation. Acquiescence to violations in the private sector is bound to permeate military sensibilities, on the one hand, and to promote acceptance of increasingly bold biotechnology activities by the Pentagon, on the other. Such conditions are sobering at face value. And they take on far greater significance in view of the DOD's predilection throughout its CBW history to bend or break rules, laws, and policies to accomplish what are considered important national security objectives.

STEMMING THE TIDE

This combination of slack regulations, conflicts of interest, and secrecy has yielded a military biotechnology program unburdened by meaningful regulation. At the very least a few stopgap measures, which would not radically alter the existing regulatory framework, can be taken immediately:

- The Goldstein-Novick proposal prohibiting development of BW by rDNA methods should be incorporated into the NIH guidelines.
- DOD adherence to the guidelines and all other biotechnology regulations should be made mandatory and carefully monitored.
- The RAC should establish a permanent working subgroup to explore the arms control implications of rDNA. This subgroup should be staffed with legal, political, and scientific experts qualified to appraise the military biological warfare defense program.
- Open-air testing of BW agents should be banned under any circumstances, consistent with the Nixon renunciation.

But in the long run the regulatory system—for both corporate and military applications of biotechnology—may require drastic reinforcement.

"Without an appropriate assessment and regulatory system, people may find the Luddite position—absolutely no tampering in the biosphere—as the only rational choice," Krimsky said about the deliberate release controversy. If anything, the point is even more relevant to biological warfare research.

9

PAYING THE PIPER: BIOLOGICAL RESEARCH AND MILITARY GOALS

THE ROLE OF THE DOD IN AMERICAN RESEARCH

Since the end of the Second World War the Defense Department has played a major, often dominant role in a broad range of scientific research conducted at U.S. universities, particularly in electronics and physics. The Office of Naval Research (ONR) and its army and air force counterparts were formed shortly after the war to maintain and expand the relationships between the military and academia that had been invaluable to the war effort. These agencies quickly became the primary conduits for federal scientific research funding.

The army and air force programs centered largely on *applied* research. But the ONR, largest of the agencies, was formed for the sake of increasing the "general fund of scientific knowledge." *Basic* research, the intellectual focus of university labs, is considered "value free." It rarely has immediately foreseeable applications. The ONR's founders assumed that a general program of basic research would best serve long-term military needs as well as preserve the military's reserve of intellectual capital. University researchers found themselves awash with funding they had never known, to continue and expand work begun before the war.

But many civilian scientists held an abiding fear of military interference. By the early 1950s other federal agencies, notably the National Science Foundation, had been established to support basic research, diluting somewhat the ONR's overall influence. The need to make DOD-funded research more immediately useful to military goals became a hotly debated topic in light of the many pressing problems of the Korean War. As a result, the ONR added a major applied research program, which was largely classified.

"Fearful that many of their best faculty would move to central government laboratories as they had during World War II, many universities decided to introduce classified research and development," explained Stanton A. Glantz, professor of medicine at the University of California at San Francisco and an expert on the history of DOD research funding. "When the Korean War ended, the fraction of the Department of Defense's research and development budget devoted to basic research increased, but the change in emphasis of the military research agencies never returned to its pre-Korean War liberality."

By 1956 the DOD had eliminated much of its general support of basic research, maintaining "fundamental research only in areas where there is a high probability of future military application," as a high-level Defense official put it.

Sputnik propelled the Soviet Union into the space age in 1957, launching a competitive frenzy. This was followed by the missile gap scare, which dominated the 1960 presidential campaign. In response, those walking the halls of government felt keen pressure to focus defense research even further. In the early 1960s Secretary of Defense Robert McNamara set into motion some pivotal changes.

"In contrast to the early postwar era, when unsolicited proposals were judged simply on scientific merit," Glantz noted, "proposals now had both to pass a scientific test and to fit into the military's research plan." This further solidified the DOD's commitment to "targeting" even so-called basic research.

The system worked relatively smoothly until the late 1960s, when the Vietnam War turned university campuses into hotbeds of revolt. Across the nation DOD contracts were targeted by activists and thrust into the vortex of protests that rocked the formerly placid academic milieus.

Against this tumultuous backdrop Senators J. William Fulbright and Mike Mansfield grew increasingly concerned that the sheer magnitude of military research spending was again becoming an unhealthy influence on the nation's scientific pursuits. Academia was growing too dependent on military contracts, they said. And some DOD contractors, such as the Hudson Institute, were publishing widely circulated position papers in support of such projects as the antiballistic missile, further angering Fulbright and other opponents of the ABM.

Meanwhile, the escalating war had become an increasingly expensive proposition, and Congress faced a budget crisis. Both to circumscribe DOD jurisdiction and to hold down overall military spending, the Mansfield Amendment was attached to the 1969 military budget authorization. It forbade spending on "any research project or study unless such project or study has a direct and apparent relationship to a military function or operation."

The House Armed Services Committee indicated how far the pendulum had swung. "The committee does not subscribe to the broad application which the DOD places upon programs and does not believe it appropriate for some of these projects to be controlled by the Department of Defense," a committee report noted. "We interpret 'military function or operation' in its narrowest sense."

Ironically, the amendment had little concrete impact on Pentagon spending because the kind of targeting that had gained ascendancy during the previous fifteen years had already brought the bulk of DOD research support into line with Mansfield's intent. (In later years, in fact, the amendment was softened.) But it threatened to have a profound effect on university researchers' psyches.

By 1969, on campuses around the country, "Opponents of DOD projects argued, on political and moral grounds, that individual responsibility required scientists to take a moral stand against U.S. policy in Indochina by refusing to work on such projects," Glantz and physicist Norman V. Albers wrote in *Science* magazine. "Backers of DOD projects argued that the DOD supported projects solely on their scientific merits and that investigators . . . were simply engaged in an unbiased search for scientific truth which happened to be funded by the DOD."

This perception reflected the open-minded philosophy of the ONR's early days, still operational in the minds of many academicians. Mansfield made this rationale untenable. Perhaps more important, the amendment elicited a telling response by the DOD. "I am going to recommend that we don't make the university scientists certify that any DOD-supported university research has a defense related outcome," Secretary of Defense Melvin R. Laird said in 1970 congressional testimony. In fact, Laird assigned this task to DOD scientific officers, effectively shielding researchers from confronting directly the philosophical and moral questions associated with their work.

STANFORD AS CASE STUDY

Despite this far-reaching controversy, very little had been done to appraise objectively the overall impact of DOD funding on the academic research environment. In 1971, however, an ambitious study conducted by a group of students at Stanford University, a major military contractor, put the issue under a microscope for the first time. Their analysis concluded that Stanford's more than 100 military contracts had severely undercut academic freedom at the university.

The students focused largely on perceptions about military research conducted on campus. First they obtained the DOD's research and technology work unit summaries, which explicitly delineate the military's goal for each contract (the same variety we used for the analysis in Chapter 6). They then showed these to professors whose projects were represented, most of whom had never seen the summaries, since they are not routinely distributed to contractors. The professors were asked to comment on the DOD descriptions of their projects. The test revealed a monumental gulf separating contractors from sponsor.

"We found that individual scientists paid with DOD money did indeed view themselves as being involved in objective searches for scientific truth," Glantz and Albers, who as graduate students were part of the research team, wrote in *Science*. They continued:

> We also found that the DOD supports research to obtain capabilities for which military planners foresee a need. . . . Since the DOD receives four to ten times as many proposals as it can fund, it merely selects those projects which fit its needs. There are nonmilitary applications of much DOD sponsored R&D but, when one assesses the nature of DOD research in the university, this random civilian "spillover" must be contrasted with the systematically organized program to develop military technology that underlies

every DOD decision to fund or not to fund a proposal. . . . The DOD considers the scientific worth of the proposals, but only after it has determined that the proposal fulfills a specific military need.

A typical example in the Stanford study was a basic research contract involving semiconductors. The work unit summary stated that the purpose of the project was to elucidate the properties of certain semiconducting materials, which would lead to their use in "night viewing devices." In striking contrast, the principal investigator called the DOD description of his research "a misstatement of the facts. . . . Absolutely no connection can be made between the studies being done here" and night viewing devices, he flatly stated.

In many cases, Glantz and Albers wrote, this difference in perception had a great deal to do with where the military saw the research fitting into an overall plan. Frequently, they said, "DOD is not concerned with the researcher's final result, but rather the process or technology he uses to achieve it. Thus, the Navy supports Stanford's superconducting accelerator program even though there are no apparent military applications for most of the physics done with the accelerator's electron beam." But to accelerate the beam, the system must be cooled to an extremely low temperature, at which its components become superconducting.

This refrigeration process, not the accelerator beam itself, was the key to navy interest. Robert A. Frosch, assistant secretary of the navy for research and development, said in 1971 budget hearings, "This large-scale refrigerator for operating at very, very low temperatures means that we can build electronic systems which will be much more compact, reliable, and efficient. The cryogenic technique would permit a great advance in certain kinds of radar and electronic warfare systems."

In a separate article Glantz wrote: "While each investigator was given a maximum of personal independence in the pursuit of his individual project, the Department of Defense had developed a highly rational structure for selecting those projects . . . that held the most reasonable promise for fitting into an overall research program. Indeed, the Department's research management structure had evolved to provide a buffer of one or two administrative layers to shelter the academic community from the military motivation for supporting the project."

There is certainly nothing sinister or even improper about the DOD's hiring researchers to accomplish military goals systematically; to do so is fully consistent with its responsibility under a congressional mandate. But to spend vast sums in a way that makes the ultimate interests of the department unknown or obscure to contractors—who have carefully been "protected" from the need to assign military objectives to their work—poses grave dangers. It can corrupt the integrity of the academic environment.

At Stanford, where defense contracts provided a majority of funding in some fields, DOD goals decidedly influenced overall research directions. "While the scientific process as reflected in each individual project proceeded objectively, funding availability biased scientists' choices on which projects to pursue," Glantz and Albers noted.

Not surprisingly, this process also had long-term effects on graduate education in the sciences supported largely by the DOD. The Stanford study found an extremely strong relationship between the percentage of DOD sponsorship in a particular university department, such as materials science or electrical engineering, and the percentage of graduates from that department who went on to work in the Defense Department or in corporations heavily dependent on defense contracts.

Glantz and Albers concluded:

> The system of DOD sponsored work at Stanford raises serious questions about the university's efforts to fulfill its role of protecting the processes by which people search for scientific truth. For non-scientific standards set outside the scientific community to have a heavy influence on the choice of which projects are undertaken may be proper and desirable for industry or government; but if one believes that universities exist in part to foster the unbiased development of human knowledge, it is not compatible with the universities' role as agency to protect the scientific process.

In recent years many of the same concerns have been renewed with increasing vigor. From 1965 to 1980 military and civilian agencies shared federal research and development spending about equally. During the Reagan administration the budget for military R&D has enjoyed a meteoric increase at the expense of all other agencies. In 1986, when its total R&D spending hit $42 billion, the DOD consumed 72 percent of all federal research and development dollars—the highest proportion in twenty-five years.

Most of the outcry over this budget has been focused on the Strategic Defense Initiative ("Star Wars") and other large weapons systems that siphon off a major pool of scientific talent and redirect its efforts away from civilian goals.

John P. Holdren, a professor at the University of California at Berkeley and former president of the Federation of American Scientists, reinvestigated the problem on a national scale in 1986. His report validated the effects of Pentagon spending on graduate education identified years earlier in the Stanford case study. More than a fourth of young scientists and engineers are now channeled to the DOD or defense contractors, Holdren concluded.

And MIT economist Lester Thurow estimates that 40 percent of the nation's engineers and scientists are linked to military efforts, at great cost to American competitiveness in the development of technologies for civilian use.

THE LIFE SCIENCES CONNECTION

The DOD administers its life sciences research in the same general manner as it does the physical sciences. The historical influences described above, and the policies that have flowed from them, affect contracts for rDNA similarly to those for X-ray laser weapons. But can today's situation in the life sciences—a field in which the DOD has always been a relatively minor player—be meaningfully compared with the obvious DOD influence in physics or engineering?

The answer depends, in part, on the military's goals. The DOD has never dominated biological research because it has had no need to do so. Even under the most perversely conspiratorial scenarios of BW research, this military mission could not rival the budget for the MX missile or Star Wars.

But beyond this obvious conclusion, how influential is military funding in the life sciences? Clearly the DOD is not a philanthropic organization. The Pentagon is unconcerned with the benefit of its funding to universities, except insofar as it promotes military goals. "The extent to which a project aids the educational function of the university is not important in the decision to grant a contract," according to DOD officials interviewed for the Stanford study. "This is presumably a consideration of the university in approving and forwarding the proposal."

Support for university purchase of expensive biotechnology equipment demonstrates this perspective. Institutions that receive these funds are logical centers for future DOD research. The University of Pittsburgh was awarded $2.24 million for a Technology Center for Biotechnology. The object of the center, according to DOD documentation, was to "provide broad guidance, research and training to augment the biotechnology program at CRDC [Chemical Research and Development Command of the U.S. Army]."

Another typical contract awarded $60,000 to the University of Washington for a neurotoxin research facility. "The instrumentation will make possible or improve research which supports the Army thrusts in CBD [chemical and biological defense] and biotechnology," according to the DOD. Other such contracts are intended to "improve the Army technology base."

As DOD funding for biotechnology becomes an increasingly large slice of the biological research pie, these kinds of contracts begin to attract schools or individual scientists that were formerly sustained by the National Institutes of Health. Projects that might have elicited a collective yawn from the scientific community years ago, such as the development of mycotoxin countermeasures, suddenly are a keen focus of interest.

The DOD is a long way from eclipsing the NIH as the dominant force in setting biological funding trends and research policies in the United States. Yet an insidious military presence in the life sciences is beginning to be felt. This liberal sprinkling of funds for research, equipment, and training is consistent with the goal of building a workable network of expertise to be used or adapted as noted in Chapter 6. In this way it is conceptually and philosophically indistinguishable from the DOD's more pervasive presence in electronics or materials science.

Of course, not all DOD biotechnology R&D is as narrowly focused as solving the riddle of yellow rain. Many studies are not obviously linked to CBW. As in other fields, the military funds a certain amount of work that could easily contribute to pressing civilian needs. Some researchers who for political or moral reasons would reject research projects oriented exclusively on CBW do not see any reason to refuse Pentagon money that contributes to broader societal goals. They argue that it is foolish to reject the funding on moral or political grounds when similar or identical studies are published in the open scientific literature, therefore are freely available to the DOD for the price of a few subscriptions.

"I'm delighted that the military is willing to pay for this research because if I can get the money from them, instead of the NIH, that saves $100,000 [of NIH funding] for another investigator," said John Baxter, an internationally known molecular biologist at the University of California at San Francisco. Baxter had a contract to clone the gene for acetylcholinesterase.

Baxter's work might find a broad range of applications—from the production of new, supertoxic chemicals to nerve agent antidotes or to important discoveries on how to cure neurological disease. "This is high-priority research we'd be doing no matter what," he added. "To take chunks of military money and divert it to health is [*sic*] good for the country."

Of course, there has been a "diversion," but not in the sense Baxter meant. The funds he used were part of the net transfer from civilian to military agencies that has taken place over the past few years. It didn't represent a genuine windfall.

The military endorses Baxter's argument that its work—from both in-house studies and contracts—produces spin-offs vital to society. As Glantz and Albers noted about physics research, however, random civilian benefits from a major biological research program targeted to specific military goals are at best a tremendously inefficient approach to pressing public health problems. At worst they are a smoke screen.

"Researchers need to consider the greater implications of what they're doing," Glantz said in a recent interview. "Implicit in the acceptance of these funds is that you're directly involved in a larger scheme that serves military ends. . . . One answer to that [concern] is intentional ignorance."

THE MANIPULATION FACTOR

Do critics like Glantz overstate the case? Or are the Baxters of the world naive? A chapter from CIA history opens a window on whether a life sciences funding program could be used for dangerous manipulation. The CIA's secret MKULTRA program to discover chemical, biological, and psychological methods of mind control was introduced in Chapter 2 and has been discussed elsewhere in this book. Even the limited records released during a 1975 congressional investigation paint an intricate and sobering picture of deliberate manipulation of the academic community.

Foundations with sympathetic executives were used to funnel MKULTRA funds for behavioral studies to academic researchers across the nation. In some cases the Office of Naval Research cooperated with the CIA by passing along funds to test hallucinogenic drugs. (The navy had by this time ended its own search—code name: Project Chatter—for a magic drug to weaken or eliminate free will.)

The agency also created its own bogus foundation, the Society for the Investigation of Human Ecology, to launder funds. The operation included an elaborate system of cover grants, which were given solely to build the society's prestige and to disguise its true interests.

Researchers from leading universities who received agency funds often had no knowledge of their true source or of CIA goals for the work. Many were appalled when they later learned their patron's identity.

Some recipients were important figures in their fields. Humanistic psychologist Carl Rogers, for example, received much of his early support from the CIA. B. F. Skinner, considered the father of behaviorist psychology, was given a $5,000 grant by the society to complete his landmark book *Beyond Freedom and Dignity*.

It may have been the best $5,000 the CIA ever spent, according to John Marks in *The Search for the "Manchurian Candidate"*:

> A [CIA] source explains that grants like these "bought legitimacy" for the Society and made the recipients "grateful." [The source] says that the money gave [CIA] employees at Human Ecology a reason to phone Skinner—or any of the other recipients—to pick his brain about a particular problem. In a similar vein, another MKULTRA man, psychologist John Gittinger, mentions the Society's relationship with Erving Goffman of the University of Pennsylvania, whom many consider the 1970s' leading sociological theorist. The Society gave him a small grant to help finish a book that would have been published anyway. As a result, Gittinger was able to spend hours talking with him. . . . Goffman unwittingly "gave us a better understanding of the techniques people use to establish phony relationships"—a subject of great interest to the CIA.

To be sure, the MKULTRA episode is considered shocking even by CIA standards. This is not to suggest that the same wholesale, systematic manipulation of the academic community is being engineered by the DOD under the guise of defensive biological research. The relative openness of today's BW program is a far cry from its counterpart before 1969, let alone the cloak-and-dagger qualities that encouraged MKULTRA abuses.

But any realistic appraisal of military BW research must take into account the numerous skeletons in the closet of the pre-1969 program, some of which were as secretive and reprehensible as MKULTRA. When examined in all its subtleties, today's BW research program offers numerous opportunities for subterfuge.

Much of the program is still kept secret. And the Pentagon may weave secret goals into even its publicly disclosed program. This work is scattered among hundreds of universities, corporations, and in-house labs. Few in the military bureaucracy—and certainly no contractors—fully understand the extent of the program or how each of its components interact. In addition, as noted in Chapter 6, there may be a pattern of relationships between these highly dispersed projects that belie offensive goals.

Similarly, the rationale for some of the core projects reviewed in Chapter 6 reflects an overriding interest in the processes or technologies used to achieve a certain research result rather than the ostensible product of the experiment. For example, in a DOD-supported study of repellents and attractants (pheromones) produced by insects at the Virginia Polytechnic Institute and State University, the approach was "to produce an insecticide-resistant strain and study its dispersal in controlled and field conditions." It would not be surprising if military interests were focused more on the dissemination of these potential biological vectors than on the chemistry of insect pheromones.

As noted above, the DOD and its contractors often have different goals for the same work. And the DOD offers a number of contracts that have clear relevance to civilian problems, which occasionally attract big-name researchers. H. Gobind Khorana, a Nobel laureate at MIT, for example, has received substantial sums to clone various visual photoreceptor proteins. And several other well-established scientists, such as Baxter, have scooped up contracts in their areas of expertise. In this way—not unlike MKULTRA funding of Skinner and Rogers—DOD interests are sufficiently expansive to confer a mantle of legitimacy on the overall program.

Harvard's Matthew Meselson points out, however, that for pragmatic reasons alone, most established biomedical researchers shun DOD funds. Top scientists have ready access to traditional sources and are unlikely to be stimulated by the comparatively narrow problems of CBW. Why move from the cutting edge of biology—the province of exciting discoveries, honors, and awards—to focus on the genetics of cobra toxin or yet another treatment for the plague? Such relatively pedestrian research—the bread and butter of the Pentagon program—is hardly of overriding importance compared with cures for AIDS or cancer.

This would seem to leave the field relatively open for scientists whose work is less successful and less respected and whose funding options are less secure. The work unit summaries we analyzed seemed consistent with this view; many of the investigators have poor publication records and pursue research more descriptive than ground-breaking, more derivative than innovative. Compelling financial needs may make these scientists unlikely to ask the military—or themselves—probing questions about the intent and context of their contracts. "Intentional ignorance" is thus a survival skill.

An unusual commentary by a high Pentagon official suggests that a troubling companion attends this financial pressure. Donald Hicks, undersecretary of defense for research and engineering, in criticizing opponents of Star Wars, stated that DOD contracts should be restricted to scientists who refrain from public criticism of the government.

"I am not particularly interested in seeing department money going to someplace where an individual is outspoken in his rejection of department aims, even for basic research," he told a Senate committee during his confirmation hearings in 1985. "I have a tough time with disloyalty," he added in an interview with *Science* magazine. "We're in a situation where we're trying to protect the position of the United States against a power that would like to soak us up. . . . Now, if someone doesn't believe that, that's his perfect right as an American citizen. I feel, if we listen to him, it would take our country down the tubes. . . . If he wants to get his money someplace else, that suits me fine. My money is overall specified to be given to people who feel the same kind of urgency that I feel."

In the fields dominated by DOD funding, of course, the suggestion that critics seek support elsewhere is fatuous. And for many scientists who are being squeezed out of life sciences research by NIH cutbacks, the comments give rise to grave concerns.

Hicks explained that this policy is unwritten and that he might fund critics on a case-by-case basis if convinced that their work were vital to DOD goals. The clarification was not particularly reassuring to the scientific community.

"I read [Hicks's] remarks with dismay. Their effect is chilling," Sidney D. Drell, president of the American Physical Society, wrote to *Science* in response. "Surely the best—indeed the only—way to nurture and sustain a community of the best scientists and engineers is to support the best work. Are we to now understand that political loyalty to DOD programs will also be a standard for deciding who and what to support?"

Hicks responded to this criticism by giving assurances that the department does not apply political "litmus tests" and that "we also try to foster an environment that encourages controversy and diverse viewpoints." He did not, however, back away from his original views.

Each of these factors—broad distribution of contracts controlled by a central authority, differing goals between researchers and sponsor, reduced resources from alternate sources, and explicit political threats—is certainly suggestive. In concert, could they help shield a broad pattern of offensive intentions?

Clearly such a plan would be extraordinarily difficult to carry off. Yet some critics of BW defense work, such as Novick of the New York Public Health Research Institute, believe it is possible. Meselson finds such a scheme implausible. But he did acknowledge that some "naive individuals" may believe "technological breakout" is feasible by obscuring nefarious goals in a plethora of ostensibly innocuous studies.

For its part, the army knows that secret intentions or information might be uncovered through the sifting of a multitude of unclassified studies. One indication is its response to our Freedom of Information Act request regarding technical data and analyses of hundreds of BW field tests conducted in the 1950s and 1960s. The army acknowledged the existence of thousands of unclassified documents relevant to the request. Yet it refused to release even an index of these aged studies on the ground that the broad programmatic "mosaic" it would reveal is classified.

"BASIC" RESEARCH?

During the early 1980s the Reagan administration conducted a major review of federal research and development. "Perhaps the most important element of policy that emerged from those reassessments was a renewed—and considerably strengthened—commitment to federal support for basic research," wrote the president's science adviser George A. Keyworth in 1984. "Not only is basic research an essential investment in the nation's long-term welfare, but it is largely a federal responsibility because its benefits are so broadly distributed. Quite simply, basic research is a vital underpinning for our national well-being."

These were welcome words to scientists. This kind of policy would mean excellence and progress for American education and technology, they reasoned. In fact, from 1981 to 1986, in a trend reminiscent of the postwar years through the early 1960s, the percentage of total federal research and development outlays spent on basic research surged.

As the new policy took hold, however, some educators grew skeptical about whether it really represented a resurgence of value-free science. "In the past [federal funds] were given with less control, fewer strings, and assurances that it would be available for long-term research," said MIT Professor Jerome B. Weisner, a science adviser to several presidents.

Other leaders in the scientific community suggest, furthermore, that the advertised increase in basic research funds has been greatly exaggerated. Instead, projects that may once have been designated "applied"—intended to result in specific findings—are now called "basic." "The camouflage is very good," noted Stuart Rice, dean of the division of physical sciences at the University of Chicago. "It may be called basic research, but the projects that get funded have applicability in the short term."

"There is a lot of room for ambiguity in the numbers," according to Susan Kemnitzer, a legislative specialist at the National Science Foundation. "When the administration said it was committed to basic research, the bureaucracy tried to jam more projects into that category."

Concomitantly, some researchers are now acknowledging that they skew their interests to attract industry or federal sponsors that have particular goals in mind. In so doing, they contribute to the erosion of basic research. These phenomena are troubling to Rice and other critics. They see the nation's future being mortgaged for short-term gains.

Although the DOD budget does not reflect an extraordinary shift to the "basic" category, ambiguity in the basic-applied dichotomy may constitute an important factor in how the DOD obtains sorely needed expertise from sometimes reluctant academicians.

Most university research funded by the Pentagon, while nominally "basic," is anything but pure, Robert Park, public affairs director of the American Physical Society, said in 1983. He cited the following basic research needs on a DOD list: self-contained munitions; electrooptic countermeasures; low-speed takeoff and landing. "I would have a hard time finding a scientist who would define these as basic," Park said. "They're funding an awful lot of applied research and masquerading it as basic."

Similar confusion appears to afflict the Pentagon's view of basic research in biology. In its fiscal year 1982 obligation report, for example, the DOD categorized as basic an "investigation of rapid biological aerosol detection based on ultraviolet fluorescence." And in the 1984 report a basic research project was "a mouse aerosol model to assess comparative toxicity of trichothecenes, time-related distribution of T-2 toxin, and toxin-induced hemostatic derangement." (The report announced that "T-2 aerosols generated from dry powder caused exposed mice to develop a dry inflammation of the conjunctiva, nose sinuses, and lips.") Clearly such investigations are far removed from the search for fundamental biological principles.

Rice believes that given its dominant research and development budget, the military greatly harms society by promoting applied research in this way. Basic research funds lost to drifting definitions are not being replaced by any other source. Although they say the evidence is largely anecdotal, Richard Atkinson, chancellor of the University of California at San Diego, and Carlos Kruytbosch, chief of the National Science Foundation's *Science Indicators* department, concur with Rice.

The military's motives are understandable. Applied research is the most direct way to accomplish important missions that the DOD is under increasing pressure to conduct expeditiously and economically. But by training and inclination, university professors, particularly in the life sciences, are devoted to basic research. Merely renaming the project "basic" allays the concerns of some. Other academicians, however, insist on unambiguously basic work.

This does not present an irresolvable problem, however, as the Stanford case indicates. "Virtually all the work could legitimately be called basic research in that it was directed at understanding fundamental properties of natural or man-made systems, as opposed to contributing explicitly to a particular weapon," according to Glantz. "However, as we shifted our focus from the individual project to groups of contracts at Stanford, overall weapons-related programs emerged." These categories of studies, unbeknownst to the researchers, proved to be wholly consistent with the DOD's practical, near-term objectives. In this way, the lines between basic and applied research are further muddled.

Among federal agencies that support biological research, only the DOD relies almost exclusively on the *contract* system rather than award funds in the form of *grants*. Contracts tend to delineate a project area rather precisely and to target specific objectives to be accomplished within a set time period. They are often solicited by a "request for proposals."

Grants, in contrast, are normally given for research within a more general area of work pertaining to the scientist's interests and expertise. There is rarely a goal in mind, other than elucidation of a particular scientific principle or biological phenomenon. In fact, review panels often treat these kinds of specific aims in grant applications as criteria for evaluating a researcher's ability to form and pursue a coherent hypothesis, not as benchmarks that must be reached before future funding can be considered.

First and foremost, the grant system seeks to support and reward the researcher who is not simply productive but also observant. There is no penalty for even radical changes in the course of a project, based on interesting, unforeseen findings. Contracts, by definition, are less flexible; they are designed to restrict intellectual options in order to secure a desired outcome within a predetermined time.

Many scientists believe these distinctions take contracts completely out of the realm of basic research. The delineation of specific objectives, even in nominally basic work, creates sometimes subtle but influential pressure to "deliver the goods." In the process, scientific objectivity, the cornerstone of basic research, is endangered. Contracts may be awarded for basic work, Meselson says, but this is clearly not the way to produce the highest-quality research.

This combination of factors—disingenuous definitions, skewed research choices, explicit targeting of basic research on a programmatic level, and the contract system—has special meaning for the BW program. As long as research is "basic"—assumed not to have immediate application—it helps to preserve an innocuous air in an otherwise controversial realm. This aids in recruiting reluctant scientists and focuses their activities on the problems of greatest military interest. It also makes the broad practical goals of the DOD's program harder to discern and evaluate—for universities and the general public alike.

THE IMPORTANCE OF QUALITY CONTROL

Beyond the question of using contracts to administer basic research, even more serious questions of quality control can be raised about Defense Department biological research, as reflected in the large number of poor-quality studies identified in Chapter 6. Military evaluation methods suggest a level of review that is in no way comparable with typical academic procedures and thus partly accounts for the poor research record.

As the chief funding agency for biological research the NIH sets national standards for peer review—a term that describes the independent evaluation of each grant application by a team of scientists chosen from among the leaders in their fields. The object of peer review is to ensure that the limited funds are spent on the most important projects conducted by the most able investigators. The system is predicated on the belief that taxpayers have a right to a good return on their investment and that scientific advances depend on supporting the best work.

The NIH peer review system is founded on ninety-three "study sections," each comprised of about fifteen experts in the field being considered, such as molecular biology or genetics. Panelists are chosen for demonstrated competence, based on their own research accomplishments, scientific honors and awards, and the quality of their publications in *refereed scientific journals.* (These publications solicit comments on each submitted manuscript from two or more independent reviewers—often direct competitors of the author.) Nearly all refereed journals reject more than 50 percent of submissions. The most prestigious publications, such as *Science* and *Nature,* have 85 to 90 percent rejection rates.

It is considered a high honor and important responsibility to be chosen for a study section. The groups are formed with a heavy emphasis on *independent,* balanced evaluation. Scientists are sought from different geographic areas, and no two scientists from the same institution may serve on the same panel. No individual may serve continuously. After a four-year term of service has expired, an interval of at least one year must take place before reappointment. The object is to ensure that the peer review system is open and is subject to the broad range of scientific opinion.

Study sections, which meet several times each year, look at five criteria within each proposal:

- Importance of the proposed research problem
- Novelty, originality, and adequacy of approach
- Training, experience, and competence or promise of the investigator
- Suitability of the facilities
- Reasonableness of the proposed budget relative to the research plan

Each member of the panel gives each proposal a numerical rating based on these criteria. The average of the scores becomes the panel rating. The panels may also recommend rejection of proposals that appear hazardous or unethical. The best proposals in a given specialty are recommended for funding.

These recommendations are then forwarded to "national advisory councils." These councils are comprised of twelve or more particularly esteemed scientists as well as lay community leaders concerned about and familiar with the programs of a particular institute within the NIH, such as the National Cancer Institute or the National Institute of Allergy and Infectious Disease. National councils advise a particular institute on how to establish spending priorities that reflect its overall mission. The councils may override study section recommendations, although in practice this rarely happens. The councils' views are then forwarded to the chief policymakers at the NIH institute in question, who in almost all cases accept the recommendations.

About 90 percent of NIH extramural awards are disbursed as grants for unsolicited proposals. This promotes relatively free development of creative ideas. No grant automatically receives renewed funding, and all evaluations consider both past work and the current proposal; thus each scientist who receives NIH funding undergoes continual and comprehensive scrutiny.

Because the NIH provides the bulk of biomedical research funding, its peer review machinery is the primary means of evaluating American biomedical research and its investigators.

The Defense Department quality control process works quite differently. Ostensibly the system strives to encourage innovation and risk taking, reflecting a DOD concern that the NIH peer review system is relatively conservative—that it tends to favor ideas justified by a strong track record but to stymie highly creative, unconventional proposals. There is no single contract review process for all the military services, although each shares the above philosophy.

The bioscience division of the Office of Naval Research awards about 200 contracts yearly. In a 1986 interview division leader Robert Newburgh described the ONR evaluation process. Proposals are funneled to one of two research divisions—molecular biology or systems biology—each of which has five scientific officers, whom Newburgh supervises.

The ONR receives two kinds of proposals, he said. "Preproposals," two to three pages in length, are evaluated by a scientific officer for military program goals but not for scientific value. The officer makes an independent decision whether to reject the preproposal out of hand or to encourage the scientist to submit a full proposal. Upon receipt of a full proposal, the scientific officer can unilaterally take one of these steps:

- Independently evaluate the proposal for military relevance and scientific merit.
- Solicit internal (ONR) advice on both relevance and merit.

- Obtain comments on scientific merit only, by mail, from ad hoc reviewers, who might be from a university or another military branch.
- Convene a review panel of eight to twelve scientists, similar to NIH-style study sections, to evaluate scientific merit only.

According to Newburgh, the ad hoc mail reviews and the internal ONR discussions are the most common methods. This discretion is the first stark contrast between military review and NIH-style review. Individual science officers who form the first line of review make some of the most important decisions at the ONR. The heavy reliance on mail reviews, as opposed to panels that actually discuss each proposal, is also antithetical to the direct debate and exchange that the NIH considers essential to adequate quality control.

If a panel is used, it is usually administered by the American Institute of Biological Sciences (AIBS), a private, nonprofit federation of scientists and biomedical societies employed by the DOD and other government agencies. Newburgh said he can veto any AIBS panel member for any reason. He cited a potential conflict of interest, however, as the only reason he normally would use this power.

The AIBS panels, comprised of five to twelve scientists, function in a manner similar to NIH study sections with one notable exception: The process is noncompetitive. Panels do not rank or score ONR proposals in any way. They merely comment on scientific merit and forward their views to the scientific officer.

Newburgh emphasized that panel recommendations are by no means binding and that this flexibility, together with the rejection of numerical ranking, encourages unconventional thinking. "We probably fund more young people or people who have a new idea," he said, "people who may have switched fields recently." These scientists are less likely to be funded by agencies that use ranking systems, he added. Under the ONR system, proposals judged essential to military goals can be supported even if they are viewed unfavorably by the panel.

The scientific officer—with or without the benefit of outside review—then makes a recommendation that works its way up the chain of command to the Program Council, the final authority. The scientific officer's recommendation is almost always approved, Newburgh said.

The Army Medical Research and Development Command (AMRDC), which oversees the AMRIID labs at Fort Detrick as well as other facilities, uses a system very similar to the ONR's. Unlike the ONR, however, the AMRDC *ranks* proposals for relevance to specific military missions and scientific merit. The relative weight of these requirements varies, according to Howard Noyes, the AMRDC's contract manager. Like the ONR, the AMRDC may accept proposals considered scientifically poor if they appear to fill compelling military needs.

Shirley Tove, chief of the biological sciences division of the Army Research Office (ARO), operates within a similar system. The principal difference is that the ARO does not use AIBS-style panels. Nor do its scientific officers have the discretionary power to dispense with all outside review. Instead, all ARO proposals are reviewed by mail by six or seven scientists affiliated with the National Research Council (NRC) of the National Academy of Sciences. These comments are then considered along with program requirements in a fashion similar to that of the ONR, and final decisions on which projects to fund are made by Tove's office.

A review of NRC panelists suggests a serious deficiency in this process. Of eleven scientists designated to evaluate molecular biology for Tove, only one has a background that indicates the proper expertise.

The military believes it has the best system to fulfill its mission. But the lack of required, stringent, independent peer review, and the systematic approval of scientifically questionable studies, raise doubts whether, in the final analysis, contractors are producing high-quality work.

In addition to its contract program, a large portion of Defense Department biological research is conducted in military labs. Charles Dasey, AMRIID spokesman, described the process the army uses to evaluate in-house studies. Review follows the chain of command, Dasey said. After receiving directives from the highest military leaders, the commander of the AMRDC develops his priorities, such as vaccines or drugs for particular threat agents. He then assigns projects to the labs under his jurisdiction, including AMRIID, according to their capabilities and budget constraints.

The army requires an annual evaluation of each study, conducted by division leaders and lab directors. In some cases the review extends farther up the chain of command. Work considered fruitful normally receives renewed support, while projects that seem unproductive may be cut off.

The AMRDC retains 100 consultants—university professors who are chosen by the directors of each lab to review informally in-house studies. For AMRIID, therefore, Colonel David Huxsoll might find experts in broad-spectrum antiviral drugs, a major interest of his lab. These independent consultants would then be approved by higher authorities. But their role is extremely limited. Called in at the discretion of the lab director, a consultant may be used monthly, yearly, or less. When the consultant's advice is solicited, it is just that—advice. Consultants have no formal power to review or evaluate in-house work.

The Army Science Board and the Defense Science Board, which, according to Dasey, consist of "well-credentialed experts," periodically review the broad AMRDC program and strategy. They have studied problems of general interest to the BW research program, such as yellow rain. These boards do not, however, engage in routine or detailed evaluation of individual studies.

In sum, the AMRDC benefits from no meaningful independent appraisal of short-range strategies or specific experiments. Perhaps more important, the chain of command system, while suitable for many military activities, poses inherent quality control pitfalls for biological research. Division supervisors, lab directors, and higher-level bureaucrats are subject to inevitable conflicts of interest. Preserving prestige or jobs, misplaced institutional loyalty, or fear of reprimand could stimulate inappropriate protection of the overall thrust and specific decisions governing their research programs. Truly independent peer review, as noted above, obviates each of these problems.

King, who has chaired the microbial physiology study section for the NIH, believes that without intensive independent scrutiny, the Pentagon is free to obscure its true goals.

"The Defense Department appears to be pursuing many narrow, applied goals that are by nature offensive, such as the genetic 'improvement' of BW agents," King says. "But to achieve political acceptability, they mask these intentions under forms of research, such as vaccine development, which *sound* defensive. This convoluted process inevitably results in poor-quality research. Moreover, these studies pollute the scientific literature by promoting false or irrelevant directions. In this way, they may serve to blur real scientific problems important to the overall welfare of society."

King sees the current BW research program as primarily concerned with recruiting scientific talent, increasing in-house capabilities in genetic engineering, and finding out what is technologically feasible. "Military planners are not particularly concerned with the highest-quality research at this stage," he says. "If and when they move on to actual weapons development, the review process will improve dramatically. They fully realize that without rigorous evaluation at that stage, their systems would fail in the field."

But NIH-style peer review, King points out, will always be resisted by the military. If implemented, it would undoubtedly uncover these kinds of contradictions.

The in-house military scientists who run the BW research program reflect King's point vividly. In the military system contract managers, lab directors, and investigators who control huge blocks of research money are pivotal to ensuring adequate program performance. From 329 work unit summaries of DOD biotechnology projects, we were able to identify ten primary contract managers and lab directors, who collectively administer tens of millions of biomedical research dollars each year.

MEDLINE, the biomedical research data base described in Chapter 6, yielded some sobering information about the credentials of these military scientists. Six, including Newburgh and Tove, have not published a single scientific paper since 1981. Only two—including Huxsoll, who holds Ph.D. and veterinary degrees—published with regularity. And even these articles appeared in obscure journals of questionable quality, such as the *Australian Journal of Experimental Biology and Medical Sciences.*

In fairness, as administrators and grant managers these scientists may not be expected to do a great deal of writing. Publication would obviously be a primary goal for principal investigators at military labs, however—particularly in light of the DOD's frequently professed openness on BW defense. In fact, only a few of the nineteen scientists we were able to identify as holding these positions produce a reasonable number of articles for top-echelon publications.

As noted in Chapter 6, the track records of many military scientists are worse than their administrators'. Among four investigators who controlled more than $7.8 million during fiscal years 1982–86, only seven papers, all in obscure journals, could be identified—a phenomenal $1.12 million per article.

Joel M. Dalrymple, as chief of the department of viral biology at AMRIID, is one of the BW program's most important biomedical researchers and has been given at least $1.8 million to conduct his work since 1982. But he is hardly prolific. Dalrymple's only publication credit since 1981 is an article about kidney disease in Sweden, published in the *Scandinavian Journal of Infectious Disease.*

Another researcher, K. W. Hedlund of the Walter Reed Army Institute of Research in Washington, D.C., has produced only four articles for mediocre journals from $3.4 million of research funding since 1982.

This lack of publication credentials shows the fundamental failure of military quality control. If these researchers were subject to NIH-style peer review, their grants would not have been renewed. The NIH system is based on publishing results to expand the base of scientific knowledge. Such unproductive scientists could never compete with the typical NIH grantee.

The *best* that can be said of this record is that the DOD fills many of its labs with incompetents. They cast a shadow on the entire BW research program by squandering badly needed biomedical research funding. But there is a more frightening explanation: These scientists may be conducting illicit, secret research.

THE BIGGER PICTURE

DOD biological research does not take place in a vacuum. According to a 1986 article in *Nature* by historian A. Hunter Dupree, strong scientific quality control in *all* defense-related decision making has foundered during the Reagan years. The most graphic example is Star Wars. Dupree wrote:

> The decision to fund the Strategic Defense Initiative on a vast scale was made without peer review, and no advice from scientists was sought or received. The whole process of research and development was aborted, with the technical choices being announced by President Reagan in advance. The American and British publics had to start paying the bill, and the scarce human resources were commandeered, without the slightest review or evaluation by the scientific community as to whether the project should be undertaken at all. This dangerous departure from post-war practice is unique.

There are signs that this kind of unilateral action may be on the ascendancy. "In my view, the problem of excessive oversight has now grown to epic proportions," DOD Undersecretary Hicks said, in response to increased congressional scrutiny of weapons programs, after a number of serious cost overruns and boondoggles came to light. To Hicks, "excessive oversight" has intimidated Pentagon engineers to the point where serious problems are hidden or high-risk innovation is shunned. The result, he claimed, is the production of many defective or obsolete weapons.

Hicks said classified R&D, which has increased several-fold from 1981 to 1986, is preferable. Program managers for secret projects "don't have to put up with all this crap, this oversight," he explained. "People who don't have to know a damn thing about it aren't [allowed] to come in and clutter it up."

A White House commission, led by electronics industry executive and former DOD official David Packard, largely concurred. In its 1986 report it urged that Pentagon weapons program managers be given greater authority.

Hicks and others who are pushing for less oversight probably don't give more than a passing thought to biology. They don't need to. Congress has generally kept its hands off BW research since Nixon's renunciation in 1969. In the meantime, views such as those of Hicks and his supporters have an insidious effect on DOD operations as a whole. And as the Pentagon gains more and more discretion over its vast research and development program, it certainly does not overlook biological warfare research, even if Congress does.

10

THE FUTURE OF GENETIC ARMS CONTROL

THE ARMS RACE IN PERSPECTIVE

"The trouble with arms control is that it is an unnatural act," Paul Warnke, former director of the Arms Control and Disarmament Agency, has said. "Even when in the national interest, it requires countries to tear down what they have built up." Just that having been done with biological and toxin weapons, the hardest part of biological arms control would appear to have been taken care of. Yet the process never ends.

Fortunately stemming an incipient genetic arms race is easier than waiting until "unnatural acts" are required on a massive scale to reverse biological warfare capabilities grown out of control. But where are the pressure points? History and current realities suggest that the key to preventing gene wars lies in American policy. The Soviet Union contributes, sometimes significantly, to biological destabilization. It is the

United States, however, that has played the dominant role in both the development and regulation of all weaponry since World War II.

From the first nuclear chain reaction to the development of the cruise missile, the United States has led the strategic arms race at every critical juncture. The Soviet Union usually catches up, but often only after a decade or more. For example, the United States flew its first supersonic bomber in 1960; the Soviet Union followed suit in 1975. The first submarine-launched ballistic missile was also built by the Americans in 1960; the Soviets built theirs in 1968. The United States had an operational antisatellite system in 1964; the Soviets not until 1981.

There are, of course, many influences on the relative standing of the superpowers in the arms race. Economic strength, cultural inclinations, political imperatives, and the power of military contractors all are high on the list. Yet the strength of the technical infrastructure—overall research and development capabilities—is arguably the most influential explanation for the pattern of U.S. pace setting.

The Soviet Union is unquestionably devoted to scientific advancement. Its accomplishments in space travel, weaponry, and many other areas show that it is one of the most technologically sophisticated nations.

This is reflected in Soviet priorities. In 1982 the Soviets conferred nearly 500,000 bachelor-level degrees in the natural sciences and engineering, compared with about 200,000 in the United States. The Soviets expended 3.7 percent of their GNP on research and development that year, while the U.S. figure was 2.1 percent.

On the other hand, the Soviet scientific community is crippled by professional rivalries. Westerners who have worked in Soviet research institutes cite not infrequent sabotage of rivals' experiments or records and falsification of data to attain promotions.

Research in the West is founded on extreme openness, considered essential to an environment in which colleagues stimulate each other's breakthroughs. In contrast, the Soviet "flow of information is constrained by restrictions on domestic and foreign travel," according to a 1986 report in the *New York Times*. "[Also] the poor telephone system allows little direct contact between researchers at different institutes."

The Soviets boast a major computer industry. But the equipment is slow and not yet widely distributed. The CIA and Defense Intelligence Agency place the Soviet computer industry eight to twelve years behind that of the United States. Computerized abstracts of some foreign journals are four years behind, according to Bruce Parrott, director of Soviet Studies at the Johns Hopkins School of Advanced International Studies in Washington. And some important Western journals, such as *Science,* are censored before distribution.

Rates of international collaboration mirror these problems. In 1982 American scientists coauthored 11,013 articles with peers from other countries. For Soviet scientists, the figure was 900. The consequence of such isolation is scientific myopia.

U.S. statements about the applications of genetic engineering to biological warfare paint a vivid picture of ominous, growing Soviet technological superiority. Nothing could be further from the truth. Any molecular biologist who has visited Soviet labs can attest to this.

Even with their relatively higher priority in engineering and natural sciences, by many measures the Soviets lag far behind in those fields. And in biotechnology the U.S. lead is much more pronounced because the Soviets have devoted fewer resources to the biological sciences. In 1982, for example, the United States granted twice as many bachelor degrees in the physical and life sciences, and mathematics, as did the Soviet Union. U.S. academic articles constitute fully 40 percent of the world total in biomedical research and are cited in worldwide scientific literature at a rate far higher than this proportion.

The Soviets, in fact, are still retooling for the genetic age. For example, British scientist Richard W. Ashford has closely followed Soviet efforts to control leishmaniasis, a disease spread by rodent parasites that causes serious skin ulcers. Western scientists are developing MCA and rDNA technologies to diagnose and treat the ailment. The Soviet strategy, Ashford says, is to control the rodent population and to use a primitive vaccine made from live, virulent parasites, which itself produces a long-lasting, disfiguring ulcer.

EVALUATING U.S. CREDIBILITY

These realities pose hard questions for the United States. The overall credibility of the world's most powerful nation—indeed, the world's only biotechnological superpower—is fundamental to charting the future of biological arms control.

Before the Nixon renunciation of biological warfare in 1969 the American CBW program had been punctuated by cavalier, dangerous research and contempt for public health, shrouded in secrecy and governed by unstated policies in direct contradiction to public positions. Since then the "defensive" BW research effort has been colored by numerous ambiguities, lapses of logic, violations of scientific regulations and standards, and outright deceptions.

If this state of affairs were attributable solely to naiveté or incompetence—and surely these are factors—it would be troubling but would still hold out hope for improvement. However, if the program is being orchestrated covertly to explore new offensive military options, the danger is far greater and more insidious. A fair evaluation must consider the American proclivity for deception and secrecy in the conduct of national affairs and military operations.

"Information that is vital to the security of this nation [is] and should remain classified," said Floyd Abrams, a leading constitutional lawyer, who pointed out that all presidents routinely make liberal, often self-serving use of secrecy. "Nonetheless, the information policies of this [Reagan] administration are radical and new. The across-the-board rejection of the values of information is unprecedented."

Indeed, Reagan has effected lasting reductions in the free flow of information. Among the more sweeping examples are the attacks on the Freedom of Information Act. Established in 1966, the act is considered by many people an emblem of American confidence in democracy. It was used to expose the My Lai massacre and illegal FBI harassment of domestic political groups. Corporations use it routinely to appraise markets or forecast trends. This book could not have been written without it.

"Contending that the FOIA had weakened law enforcement and intelligence agencies and become burdensome to implement," Abrams noted, "the administration made enactment of major amendments limiting the scope of the act a matter of high priority."

At first this took the form of legislative initiatives, some of which have been implemented by Congress. Most notably CIA and FBI operations files—involving intelligence methods and sources—have largely been exempted from the FOIA's jurisdiction. In the face of congressional opposition to more sweeping revisions, though, the Reagan team used its control of agency rule making to tighten guidelines for releasing information. It barred the waiver of high fees except for requests that met an *agency's* definition of public relevance, which may differ markedly from a requester's view, especially if the request relates to agency malfeasance.

Other changes have contributed to an increasingly classified society.

In 1978 President Jimmy Carter signed an executive order that required documents to be classified only on the basis of "identifiable" potential damage to national security and stated that the public's right to know was to be considered in decisions to classify documents. If classification level was in doubt, the order said, a lower designation should be used. Yet in 1981, when the federal Information Security Oversight Office conducted a random check of classification procedures, it found that in the previous year alone 800,000 pieces of information had unjustifiably been declared secret. In millions of cases unauthorized personnel had imposed security designations.

Still, in 1982 Reagan reversed each Carter reform with his own executive order.

Attempts to limit access to both government and private informational computer banks, discussed in Chapter 6, have been part of a larger strategy to reign in the scientific community. Particularly blatant episodes occurred in 1982 and 1985, when the Pentagon required conferences of photooptical engineers to cancel the presentation of more than 100 *unclassified* papers. The DOD supports much of the laser and satellite surveillance research conducted by such scientists.

The Reagan administration has also moved to silence potential leaks by former government officials. The CIA has long required its employees to submit to lifetime prepublication review of writings—from fiction to newspaper columns. This allows the agency to excise statements it deems to reveal classified information. The administration expanded that rule to cover a wide range of government and private industry personnel who have access to certain classified information. By the end of 1983, according to the General Accounting Office, an unprecedented total of 225,000 people had agreed to lifetime censorship.

The administration justifies these and other means of constricting the flow of public information as necessary and prudent in a dangerous era. But their price is high. "The system that produced the scientific and technological lead that the Government is now hoping to protect has been a basically open one," commented Abrams. "By threatening the openness of the process by which ideas are freely exchanged, the administration threatens national security itself."

THE SECRET WAR FACTOR

Covert actions have been a feature of American foreign policy for many years. In the 1970s a series of intelligence scandals revealed some of the details of this shadowy policy for the first time. While the CIA violated human and constitutional rights with drug tests and other schemes at home, it organized secret armies, assassinated foreign leaders, and overthrew democratically elected governments abroad. When these affairs were splashed across the front pages of America, stunned legislators toyed with putting the agency on a short leash, forbidding all covert action.

Cooler heads prevailed. Congress realized that it did not want to eliminate covert actions; it just wanted enough control to prevent the most egregious violations of American ideals and to bolster foreign policy goals. The legislators struck what amounted to an elaborate deal with the CIA. Never again, congressional leaders hoped, would the government find itself in the uncomfortable position of cleaning up after actions that caused acute embarrassment domestically and weakened U.S. prestige internationally. The agreement ran as follows:

- Any secret action must be consistent with publicly espoused, congressionally approved policy.
- The executive branch must fully debate covert actions before they take place.
- Congressional intelligence committees must be given advance notice of any covert action.
- The "plausible deniability" factor must be dropped.

"[The last] practice was revealed when the Senate Intelligence Committee tried to determine if any president had ordered an assassination," explained former high-level DOD official Morton H. Halperin. "It came up against a set of procedures that were designed precisely to obfuscate the degree of the president's involvement in any intelligence operation, and thus allow him to claim plausibly that he had not approved of it."

The president was allowed, however, to keep a covert operation secret for a reasonable period as long as he informed Congress in a "timely" fashion.

Public sentiment against covert actions widely considered improper was so strong that in the late 1970s the agency's roster of covert agents was sharply curtailed by Carter's CIA director, Stansfield Turner. Conservative politicians and agency insiders thought that Turner went too far, effectively eliminating what was euphemistically called the agency's "special projects" capability. When the Reagan administration took over, these critics found a home.

Under William Casey the CIA renewed its covert actions with a vengeance, in hot spots from Angola to El Salvador. "If covert operations grew any faster, they'd be listed on the New York Stock Exchange," commented Richard N. Haass, who left the State Department in 1985. Then, in 1986, the Iran-contra scandal broke.

By selling arms to Iran in exchange for American hostages, the administration contravened its own counterterrorism policy. By diverting profits from the sales to the Nicaraguan contra rebels, it violated a congressional ban on such aid. In each case it systematically and unambiguously violated each point of the covert action compromise of the 1970s.

DISINFORMATION AND U.S. CREDIBILITY

U.S. presidents have accused the Soviet Union of routinely spreading disinformation—deliberately deceptive or false news and information—for many years. There can be little doubt that the Soviets engage in this practice. Conventional wisdom has held that the United States, however, is above such actions.

A recent affair, however, offered a dramatic public contradiction of this perception. In October 1986 *Washington Post* editor Bob Woodward uncovered a memo written by National Security Adviser John Poindexter about a plan to undermine Libyan leader Muammar Qaddafi. "One of the key elements of the new strategy," it stated, "is that it combines real and illusory events—through a disinformation program."

Poindexter, who later resigned under pressure after his involvement in the Iran-contra affair had come to light, implemented this policy by planting stories in the *Wall Street Journal* and other papers that falsely suggested Qaddafi was planning new terrorist actions and was in imminent danger of being overthrown by internal opponents. This technique was later applauded as legitimate "psychological warfare" by Secretary of State George Shultz.

"In considering 'disinformation' as a means to undermine the Libyan leader, the Reagan administration has not only risked damage to its credibility but also cast doubt over its overall news policy," according to *Times* analyst Bernard Weintraub.

The *Columbia Journalism Review* cited another incident that underscores the full implications of the issue. It involved a March 1986 article published in *Aviation Week & Space Technology,* a journal closely followed by the Soviets. The article reported that "the Department of Defense and the CIA had initiated a disinformation program designed to mislead the Soviets by releasing false, incomplete, and misleading information about aircraft- and weapons-development projects."

The reporter later conceded that "material leaked to him about this disinformation campaign may itself have been disinformation," the *Review* added. "The thought is frightening, conjuring up a world in which government-authorized fictions appear as fact and every account, no matter how prestigious the publication in which it appears, is suspect."

Together, unprecedented attacks on freedom of information, increased secrecy, proliferation of covert actions, and blatant disinformation campaigns cast a long shadow. Even the most sympathetic observer of the U.S. biological warfare program has reason to view the American story with considerable uncertainty, if not with open suspicion.

CREDIBILITY AND THE "SOVIET THREAT"

"Let's not delude ourselves," Ronald Reagan said in 1981. "The Soviet Union underlies all the unrest that is going on. If they weren't engaged in this game of dominoes, there wouldn't be any hotspots in the world."

Since that statement the Pentagon has regularly pronounced the Soviet Union the leader in nuclear arms as well as in conventional forces. For the first time an American president has taken the position that Russia—the "evil empire" in Reagan's parlance—is the foremost military power. The reason for the Soviet buildup? "World domination, it's just that simple," according to Defense Secretary Caspar Weinberger.

These statements represent the leading edge of a view that enjoys substantial support among both politicians and citizens. The American BW program is founded upon and propelled by such beliefs. The "Soviet threat" and its influence on the arms race are complicated issues that involves far more than comparative statistics. In many ways, the threat is a state of mind. "Almost every new weapon has been presented as a defensive measure but perceived as an offense one," says Tom Gervasi, director of the New York-based Center for Military Research and Analysis.

As shown in earlier chapters, the administration has offered only the flimsiest evidence to support allegations of a Soviet BW buildup employing biotechnology. But if it can be shown that the *overall* Soviet threat is real and imminent, U.S. fears about the Soviet *BW* threat—however frail their apparent justification—would gain considerable currency.

In defining nuclear superiority, the Reagan administration often cites the explosive force of nuclear warheads. Soviet missiles tend to pack more megatons—equivalent to 1 million tons of TNT—than U.S. weapons. Administration officials have frequently implied that a higher aggregate megatonnage is a critical element in the nuclear arms race. This conclusion is fallacious.

"Weapons of higher yield merely provide an excess of destructive capability," explained Gervasi in his book *The Myth of Soviet Military Supremacy,* perhaps the authoritative work on the issue. "The higher yield . . . was simply an attempt to compensate for their inferior accuracy."

"In megatonnage, the Soviets are way ahead," Paul Warnke conceded during a 1982 speech. But he added that "it doesn't make a damn bit of difference" because the United States has many more warheads.

This kind of distortion, which portrays certain relatively meaningless features as decisive, is typical of U.S. arguments. In 1980 candidate Reagan attacked President Carter by claiming that Soviet SS-18 and SS-19 missiles were superaccurate "first strike" weapons. Carter had allowed a "window of vulnerability" to open, because American missiles were in danger of being destroyed in their silos. After Reagan was elected and inaugurated, his administration used this argument repeatedly in its admonitions that the United States was trailing far behind and that a major U.S. nuclear buildup was urgently needed.

In 1986 the CIA presented an appraisal considered heretical by the DOD and White House. It strongly disputed estimates of the accuracy of Soviet SS-18 and SS-19 missiles. This was one of a series of CIA statements that have questioned fundamental elements of the U.S. depiction of the Soviet threat.

The most shocking of these involved spending. Reagan's military budget increases, featuring procurement boosts averaging 13 percent a year from 1981 to 1985, were a response to massive Soviet spending increases over a decade, as reported by the CIA and the DOD. In 1986, however, after obtaining new intelligence data, the CIA changed its tune. The Soviet defense budget had grown by only 2 percent a year from 1975 to 1984, according to the revised view. Soviet weapons procurement rose an average of only 1 percent per year—approximately the rate of Soviet inflation.

"A proper accounting and interpretation of both sides' military outlays indicate that, in fact, the West outspent the Eastern bloc by $740 billion from 1971 through 1980," concluded Tufts University Soviet specialist Franklyn D. Holzman, author of *Financial Checks on Soviet Defense Expenditures*. In essence, there was never a demonstrable justification for the U.S. buildup.

Thus, even the CIA disputes bedrock tenets of the Soviet threat argument despite Reagan's persistent marketing of the notion to the American people. At the very least the most pessimistic views of overall Soviet intentions and capabilities lack sufficient support to justify the worst fears about the Soviet BW program—in the absence of solid evidence of impropriety.

"U.S. perceptions of hostile Soviet intentions have increased, not when the Russians have become more aggressive or militaristic," according to Queens College historian Alan Wolfe, "but when certain constellations of political forces have come together within the United States to force the question of the Soviet Threat onto the American political agenda." These forces include the dynamics of electoral campaigns, rivalries between branches of government, competition among the military services for power and prestige, and disputes within the ruling elite over economic and foreign policy.

Periods characterized by the most extreme views of the Soviet threat tend to follow what are seen as authoritative reports showing an imposing new aspect of the Kremlin's military power. These appraisals are always fabrications or exaggerations. The resulting fears, promoted by competing military branches, are relieved by hefty budget hikes. In the 1950s a "bomber gap" was the key comparison. John F. Kennedy rode a "missile gap" to the presidency in 1960. Ronald Reagan used the "spending gap" to decisive advantage in the early 1980s.

"Yet each new predicted emergency never arrived," Gervasi commented. "Each mistake was always officially acknowledged, though only years after it was made. It turned out, each time, that there had never been a bomber gap or a missile gap or any other gap. Each new expansion of American military power to close these gaps had therefore been unwarranted. It had always been impossible, however, to appreciate this fact until long after each expansion had already taken place."

When this record of false alarms was to question the Reagan military buildup, the questioners were often labeled agents or victims of Soviet disinformation. The nuclear freeze movement, which seeks to hold superpower arsenals at current levels, was so targeted during the military debates of the early 1980s, when the DOD budget saw its steepest increases.

So far few have even questioned the administration's depiction of the Soviet threat of genetically engineered BW.

THE TREATY RECORD

Beside the overall Soviet threat, the most influential factor affecting the U.S. position on BW is Soviet flouting of arms control treaties. As detailed earlier, the United States has accused the Soviets of multiple, flagrant violations of the 1925 Geneva Protocol (which bans first use of CBW) as well as of the 1972 BW Convention. As covered in Chapter 3, these allegations were presented with a paucity of convincing evidence. Each charge has been discounted by a wide range of independent scientists and by other governments, including U.S. allies.

Just as genuine evidence of *overall* Soviet military superiority would have served to support even weak indications of a Soviet *BW* threat, genuine evidence of overall Soviet disregard for the rule of international law could validate otherwise unconvincing claims that the Soviets are violating BW treaties.

Since the early 1980s U.S. officials have accused the Soviets of breaching a wide variety of arms pacts, including the 1972 ABM treaty, the two Strategic Arms Limitation Talks treaties (SALT I and II, signed in 1972 and 1979 respectively), the 1975 Helsinki accords on conventional force movements, and the 1974 Threshold Test Ban Treaty, which limits the size of underground nuclear explosions. As with the CBW allegations, in each case the statements have been strongly denied by the Russians and disputed by U.S. allies, independent arms control analysts, and officials from previous administrations.

Those who are best equipped to judge the accusations have objected most strongly to them. These include Warnke, who negotiated SALT II;

Gerard Smith, who negotiated SALT I under President Nixon; Herbert Scoville, a CIA official and disarmament expert in the Kennedy and Johnson administrations; McGeorge Bundy, national security adviser to Kennedy and Johnson; Harold Brown, secretary of defense under Carter; and a host of Democratic and Republican congressional leaders.

In 1987 a panel of experts convened by Stanford University released a highly detailed eighteen-month study of the issue. It concluded that "overall, U.S. and Soviet compliance with the terms of existing arms control agreements has been good." The panel warned, however, that allegations of Soviet violations are causing "a real crisis based on false perceptions." Signers of the report included former U.S. members of the committee that consults with the Soviets on possible violations and a former general counsel to the U.S. Arms Control and Disarmament Agency.

Critics point to serious inconsistencies in administration arguments. For example, U.S. officials have often called the SALT treaties "unverifiable" and therefore dangerous false advertisements that these arms are effectively contained. "The ability of the [sophisticated satellite and radar networks] to spot all of the Soviet Union's violations and provide detailed descriptions of the Kremlin's weaponry for the Defense Department's annually revised catalogs on the one hand, while at the same time being held as inadequate to the task of verifying arms control agreements on the other, remains one of the abiding contradictions of the Reagan presidency," William E. Burrows has pointed out in his book *Deep Black: Space Espionage and National Security.*

Moreover, sharp, sometimes bitter debates about Soviet compliance have arisen even within the administration and the Defense Department itself. The DOD has consistently opposed SALT II, maintaining that the Soviets have a policy of treaty violation. Yet the Joint Chiefs of Staff in 1986 released intelligence data that belied claims by Reagan and Weinberger that the Soviets had violated SALT II missile ceilings.

The State Department, which believes the Soviets have been generally faithful to the treaties, has supported SALT II. Lieutenant General John T. Chain, then director of politico-military affairs for the State Department, told a congressional committee in 1986: "If you take the body of treaties in the macrosense [the Soviets] have complied with the large majority. . . . I would hate to see [anybody] walk out of here at the end of the day thinking that the Soviets always cheat," he added. "It is certainly not the position of the State Department."

What this dissension "suggests, ever more inexorably, is a nonnegotiating strategy, pushed by officials who regard arms control as a Soviet trick to bring about unilateral American disarmament," according to a *New York Times* editorial. "Americans will find it increasingly hard to believe that President Reagan really wants to control nuclear weapons with the Soviet Union when on this issue his own government spins on, out of control."

This conclusion appeared to be confirmed in November 1986, when the administration exceeded the SALT II numerical limits in response to the claimed Soviet breach of the accord. French, West German, and Belgian officials immediately criticized the action. Within three weeks majorities of both houses of Congress had urged Reagan to return to observance of the treaty.

Here was the climax of years of efforts by the Pentagon, backed by certain administration hard-liners, to force the issue of Soviet "cheating" into the public eye, rather than hammer out problems at the negotiating table. It fueled already substantial speculation that the Pentagon was looking for ways to cancel treaties it had never wanted in the first place.

In a move that sparked an even greater furor at home and abroad, Reagan unilaterally "reinterpreted" the 1972 ABM treaty at the urging of Pentagon officials. The pact, which is of indefinite duration, is based on the idea that by restricting defenses against ballistic missiles, the superpowers would not be inclined to produce ever-larger numbers of missiles in order to overwhelm defensive systems.

Until Reagan's move in 1986 both the Soviet Union and the United States accepted what is considered the "traditional interpretation" of the treaty, which forbids any mobile land-based defense systems, as well as any that are sea- or space-based, regardless of whether such systems existed when the treaty was signed.

The new interpretation holds that only systems in existence in 1972 are covered, effectively eviscerating the pact. It permits the United States to develop, test, and deploy futuristic technologies, such as the X-ray lasers central to Star Wars. Weinberger announced in early 1987 that under this new view initial Star Wars deployment will begin in 1994.

The reinterpretation ignored objections by almost everybody, including Congress and European leaders, such as the normally dependable Reagan ally Helmut Kohl, the West German chancellor. It ignored a protest by six former secretaries of defense from both Republican and Democratic administrations. All but one of the senior negotiators of the ABM treaty disapproved of the new interpretation. Even the exception, hard-line cold warrior Paul Nitze, espoused the traditional interpretation until a year earlier.

New York Times columnist Tom Wicker wrote that the novel analysis makes the treaty "a dead letter." He called the move a "self-serving interpretation that clever lawyers have tortured out of what up to now has been considered the treaty's unambiguous language." Failing to abide by the treaty's restrictions in this manner, he added, would be a bellicose act of bad faith and would virtually assure that the Soviets would follow suit.

Moreover, with the exception of progress toward eliminating medium-range nuclear missiles from Europe, U.S. arms control policies of the 1980s are widely considered provocative, making those policies profoundly destabilizing. The nation has traveled a long distance in a dangerous direction. "During the Reagan years, the United States has generally downgraded its commitment to international law and internationalism in a number of different domains," said Princeton's Richard Falk.

Beyond actions related to the SALT and ABM pacts, he cites U.S. rejection of the Law of the Seas Treaty, repudiation of acceptance of the compulsory jurisdiction of the International Court of Justice (World Court), boycott of that court's proceedings initiated against the United States by Nicaragua, withdrawal from UNESCO, and rejections of the 1977 Protocols to the Geneva Convention on the Law of War.

"Such a posture emphasizes unilateral action, military capabilities, all-out conflict, and tends to diminish the worth of normative efforts to restrain undesirable patterns of behavior," according to Falk. "If restraints on a specific type of warfare or weapon between the superpowers erode, then prohibited forms of behavior are likely to spread rapidly throughout the state system as a whole, especially if the type of warfare and weaponry involved is expensive and possesses strategic advantages."

A greater danger begins when U.S. leaders start to believe their own public distortions of the Soviet character, Gervasi wrote. "Then they will lose their ability to see events as the Soviets see them. They will no longer be able to predict how the Soviet leadership may respond. Then it will be the United States, not the Soviet Union, which overreacts and risks human extinction."

LIVING IN THE REAL WORLD

For all its shortcomings the Reagan worldview incorporates an important truth: We live in perilous times. The genetic manipulation of biological weapons agents is a frightening reality. A government that did not develop some kind of response would be abdicating its responsibilities. But major changes are warranted in how the United States approaches the problem of biological warfare. These changes must address some hard realities:

- While there may always be some temptation to use biological weapons under limited circumstances in the third world, they are very unlikely ever to become a significant threat against developed nations.
- A research and development program that claims to focus on medical defense is a fundamental misrepresentation that inevitably obscures or leads to offensive goals. No technological breakthrough can change this fact.
- Strong international agreements are vital to containing the biological arms race. The 1972 BW Convention and the agreements reached at its 1986 review conference are essential, hopeful steps. But they must be supplemented by vigilant, independent scrutiny of military activities to ensure faithfulness to international law and to curb borderline research that may ignite a torrent of charges and countercharges.

On the basis of these points the following actions would reduce the U.S. role—intentional or inadvertent—in the escalation of biological warfare activities.

Step 1: Eliminate all secret research and development associated in any way with biological warfare. The United States would lose nothing by revealing equipment and "threat assessment" studies, the only BW data now officially classified. If anything, it would gain international prestige for openness. The utility of BW would be reduced as other nations benefited from knowledge of American defensive technology. The government would well serve its stated goals by laying to rest the doubts about its openness in areas that are called "unclassified."

Step 2: Eliminate the program of medical defense against BW agents. The Defense Department is not equipped to conduct the highest-quality biological research for any purpose—be it BW defense or the tackling of endemic diseases encountered by soldiers. Given its unsavory record over the decades and its vague, misguided notions about the use of biotechnology, the Pentagon can by no means be trusted to conduct hundreds of millions of dollars' worth of biomedical research each year.

The only exception should be the development and production of garments and vehicles impervious to chemical, biological, and toxin agents. This requires no testing with pathogenic organisms or toxins. Ultimately this work should be internationalized to ensure that all nations have access to the highest-quality protective technology. In this way the prospects of an overt CBW attack on any nation would be dramatically reduced.

Step 3: All remaining biological research funds should be placed under the jurisdiction of the National Institutes of Health. The NIH could be directed to designate a reasonable level of funds for legitimate military medical problems not linked to BW defense. In this way, rigorous, comprehensive peer review would replace the military's idiosyncratic evaluation methods, which neither serve genuinely defensive goals nor get the most out of scarce biological research funding.

These steps would in no way harm national security. To the contrary, if it followed a true policy of openness, the United States would promote stability by offering a tangible guarantee that its work is not offensive. By deemphasizing the importance of BW, the United States would promote the reality that such weapons are of relatively trivial significance—except to turn small conflicts into major conflagrations or to cause economic, public health, and ecological disasters. By charging NIH study sections to oversee military medical needs, the United States would vastly reduce these threats and any prospect of nefarious use of the new genetic technologies.

Only a strong combination of international and domestic pressure can bring about this sweeping redirection of energies. The new genetic technologies are poorly understood and often mistrusted by the general public. To progress toward the above goals, individual scientists, their organizations, and the biotechnology industry must take leading roles in public education and congressional lobbying. They must become translators not only of the breakthroughs of their work but of its potential dangers.

Unlike the link with engineers or physicists, the historic link between biologists and weapons research is weak. Indeed, biologists have played a pivotal role in the control of BW. They helped design and lobbied for the 1972 BW Convention and for ratification of the 1925 Geneva Protocol. The American Public Health Association has been at the forefront of professional pressure for biological and chemical disarmament for decades.

The Boston-based Committee for Responsible Genetics (and its affiliate, the Committee on the Military Use of Biological Research) occupy the liberal wing of the genetic research community. Its members include a number of eminent molecular biologists as well as historians, environmentalists, and labor and public health leaders. These groups are organizing opposition to military use of biotechnology within the scientific community, including circulating a petition that states opposition to any biological research that could threaten the treaty regime. This small coalition effort, however, is exceptional.

Most molecular biologists shun contact with social activists and have had little to say about biological weapons in the age of biotechnology. "Why is it that scientists look upon opposition to weapons as 'political,' but work on the development of weapons as 'patriotic'?" asked a delegate to the 1986 review conference of the BW treaty. Industry groups and professional organizations have lost the moral high ground to critics such as rDNA gadfly Jeremy Rifkin. His lawsuits against military biotechnology have often filled a vacuum left by scientists' inaction.

Ironically, some of this torpor is a response to the probing moral, ecological, and social questions Rifkin and others have raised about the technology itself, regardless of its military applications. Rifkin is anathema to molecular biologists. They reject him as a naive interloper and fearmonger when his questions touch biomedical research funded by civilian agencies or corporations. Yet they have ceded their responsibility to him when it comes to military biotechnology. By assuming a defensive posture and failing to articulate their positions effectively to the general public, scientists risk squandering public support and have heightened mistrust of genetic engineering.

More important, while Rifkin's legal and confrontational tactics can cause the military considerable inconvenience, only scientists have the prestige and credibility to redirect the energies and attitudes of their own community. By largely skirting the controversy swirling around biological warfare, scientists miss an important opportunity to assume a leadership role against the misuse of biotechnology for military purposes. In this way they could show the public where their ultimate commitment lies: with the welfare of society and the peaceful use of genetic engineering.

At the very least the National Academy of Sciences and the American Society for Microbiology should establish official oversight committees to monitor their own work and the DOD's—to ensure that researchers scrupulously respect the BW treaty. It should be these groups, which are pillars of the scientific establishment, that organize conferences and circulate petitions against the military use of biotechnology. Ultimately scientific societies are only one of the power centers that will influence the direction of military biology. But they have a critical role to play in educating the public so that the level of debate on these issues rises above confusion and ignorance.

But more than education and moral suasion are required. Professional societies that include life scientists should amend their charters to charge each member to reject work on biological warfare, even if it falls under the rubric of "defense." Offenders should be expelled. These groups also have an obligation to urge their counterparts in other nations, including the Soviet Union, and the international societies of which they are members to adopt similar policies.

These steps alone cannot solve the problem of biological warfare, of course. Improved superpower relations and general progress on chemical and nuclear disarmament are inextricably linked to that goal. But they are necessary steps that could help ensure that the ultimate life science will be used exclusively for human benefit.

APPENDIX 1

PREPARING RECOMBINANT DNA

To prepare recombinant deoxyribonucleic acid (rDNA), scientists usually use bacterial *plasmids*—double-stranded pieces of DNA formed into free-floating loops that are minute compared with the very long DNA strands of *chromosomes* that constitute each individual's genetic material. By use of a centrifuge, plasmids are easily separated from chromosomal DNA and then mixed with *restriction enzymes*. These specialized proteins cleave the plasmids with surgical precision at specific sites. The enzymes also leave "mortised" ends on each cut loop—formed by a small piece of a single strand of DNA. These mortised ends enable the plasmid to be attached to other genetic material as shown in Figure 1a.

The same restriction enzymes are then used to clip out a single gene from the DNA of another species, say, a virus. The "donor" virus gene is then inserted, or "spliced," into the gap in the plasmid loop, taking advantage of its mortised ends. The result: a chimeric molecule containing the DNA from two species. In this way any gene from any species can be spliced into a plasmid.

FIGURE 1A
PREPARING RECOMBINANT DNA

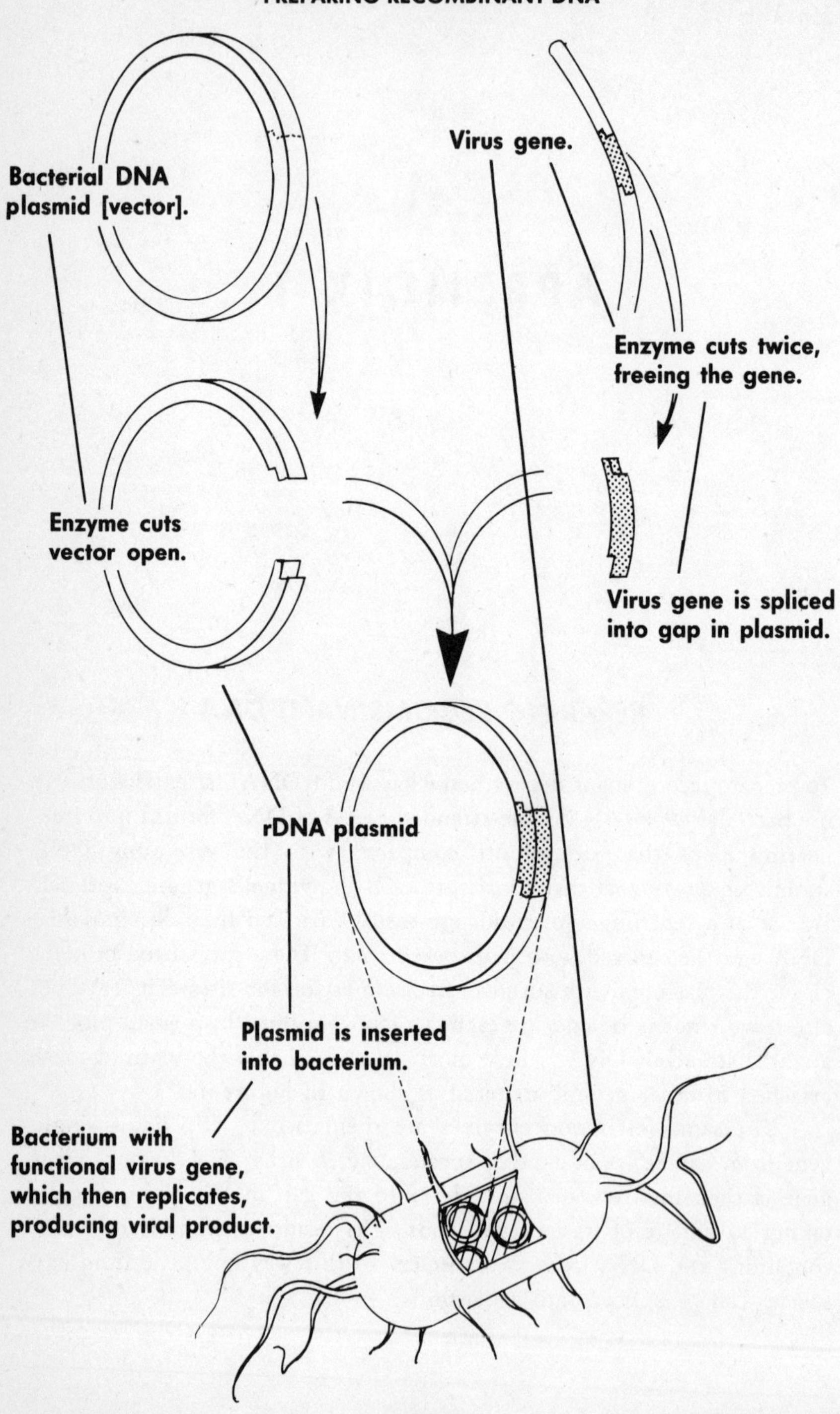

FIGURE 2A
PRODUCING MONOCLONAL ANTIBODIES

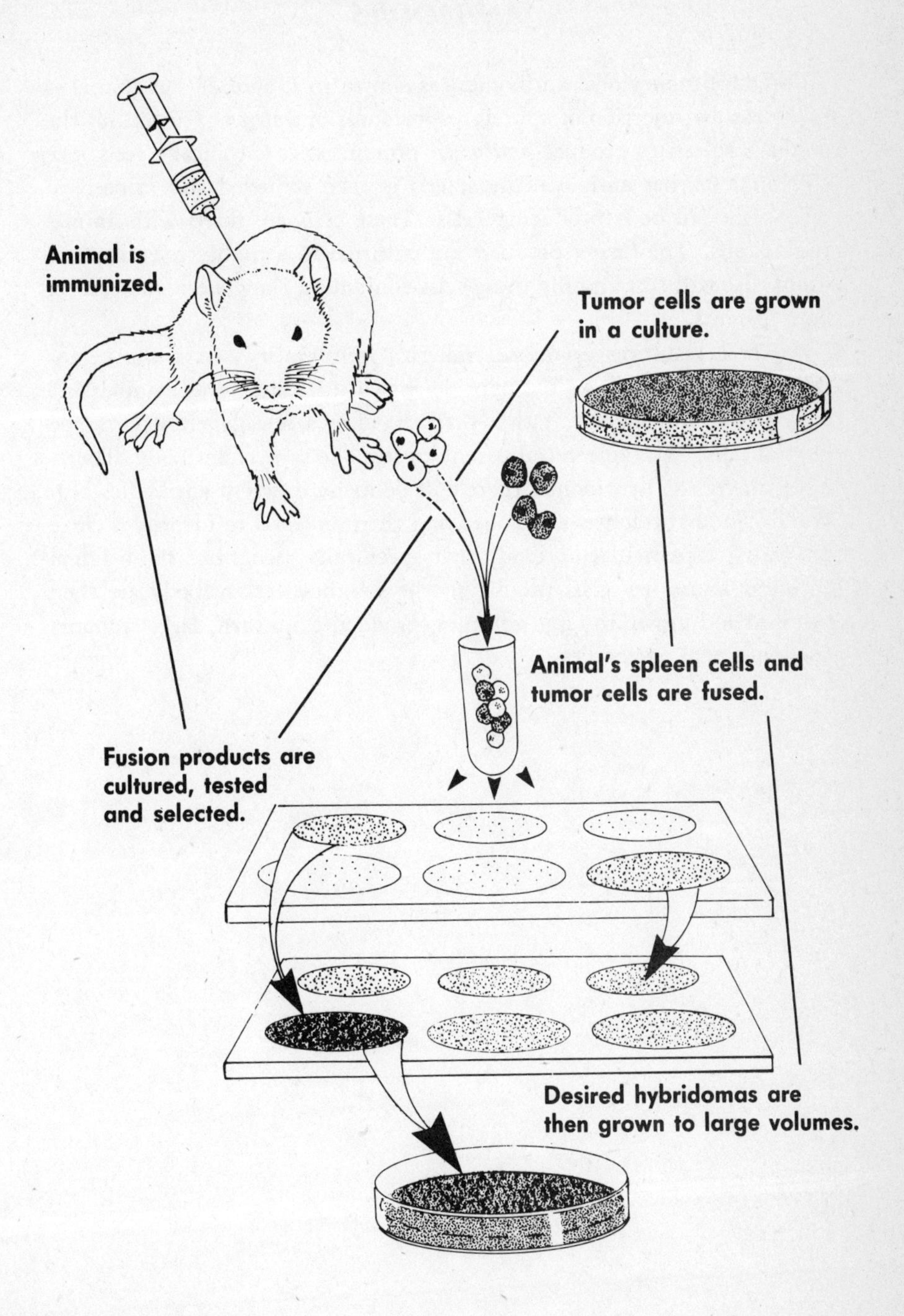

PREPARING MONOCLONAL ANTIBODIES

To produce monoclonal antibodies, as shown in Figure 2a, an animal is *immunized* by injection of a foreign substance, or *antigen*. This causes the animal's spleen to produce *antibodies,* proteins that specifically recognize and bind to that antigen. The spleen is then removed and minced to release the antibody-producing cells. These cells are fused with animal tumor cells. The fusion products are cultured in a highly selective medium that will sustain only those cells containing the genetic material of both parents.

These resulting *hybridomas* inherit "immortality" from the tumor cell parents. This allows the hybridomas to replicate—and manufacture antibodies—indefinitely, rather than die after a few generations, as do normal cells. Each spleen cell parent creates one type of antibody. Therefore, among the hybridomas there will be many different antibodies produced. The hybridomas are cloned and then screened to identify a clone producing a particular antibody that specifically recognizes the original antigen. Those few cells producing the sought-after antibody are then recloned and grown to large volumes, producing, in turn, large amounts of monoclonal antibodies.

APPENDIX 2

TABLE 1A

PATHOGENIC MICROORGANISMS STUDIED AS POTENTIAL BW AGENTS*

NAME OF DISEASE	CAUSATIVE AGENT	DEATH RATE (%) IN UNTREATED CASES OF NATURAL DISEASE	LIKELY MODE OF DISSEMINATION
	ANTIPERSONNEL AGENTS		
Viruses			
Influenza	IV	0–1	Aerosol
Psittacosis	*Chlamydia psittaci*	10–100	Aerosol
Russian spring-summer encephalitis	RSSEV	0–30	Aerosol or tick vectors
Yellow fever†	YFV	4–100	Aerosol or arthropod vectors
Dengue fever	DFV	0–1	"
Chikungunya	CV	0–1	"

NAME OF DISEASE	CAUSATIVE AGENT	DEATH RATE (%) IN UNTREATED CASES OF NATURAL DISEASE	LIKELY MODE OF DISSEMINATION
Venezuelan equine encephalomyelitis†	VEEV	0–2	"
Rift Valley Fever	RVFV	0–1	"
Eastern encephalitis	EEV	50	"
Western encephalitis	WEV	0–3	"
Japanese encephalitis	JEV	2–50	"
St. Louis	SLEV	2–22	"
Junin (Argentine hemorrhagic fever)	AHFV	5–15	"
Lassa fever	LFV	1–50	Aerosol; arthropod or rat vectors
Marburg fever	MFV	35	Aerosol
Smallpox	Variola virus	10–30	"
Ebola	EV	65–80	
Hepatitis A	HAV	Low	Multiple routes
Bacteria			
Tetanus	*Clostridium tetani*	90–100	"
Gas gangrene†	*Clostridium perfringens*	High	"
Plague†	*Pasteurella pestis*	30–100	Aerosol; flea/rat vectors
Anthrax†	*Bacillus anthracis*	95–100	Aerosol
Glanders	*Actinobacillus mallei*	90–100 90–100	" "
Melioidosis	*Pseudomonas pseudomallei*	95–100	"
Cholera	*Vibrio comma*	10–80	Water contamination
Typhoid	*Salmonella typhosa*	4–20	Aerosol; water/food contamination

NAME OF DISEASE	CAUSATIVE AGENT	DEATH RATE (%) IN UNTREATED CASES OF NATURAL DISEASE	LIKELY MODE OF DISSEMINATION
Dysentery	*Shigella* species	2–20	Water/food contamination
Tularemia†	*Francisella tularensis*	0–60	Aerosol
Brucellosis	*Brucella* species	2–5	"
Rickettsiae			
Epidemic typhus	*Rickettsia prowazekii*	10–40	"
Q fever†	*Coxiella burnetii*	1–4	"
Rocky Mountain spotted fever	*Rickettsia rickettsii*	10–30	Aerosol; tick vectors
Fungi			
Coccidioidomycosis (valley fever)	*Coccidioides immitis*	0–50	Aerosol

ANTIANIMAL AGENTS

NAME OF DISEASE	CAUSATIVE AGENT	DEATH RATE (%) IN UNTREATED CASES OF NATURAL DISEASE	LIKELY MODE OF DISSEMINATION
Viruses			
Foot-and-mouth disease (cattle)	FMDV	3–85	Aerosol; water/food contamination
Rinderpest (cattle plague)		15–95	"
Vesicular stomatitis (cattle)		15–95	"
Newcastle disease (poultry)		10–100	"
Fowl plague		90–100	"
African swine fever		95–100	"
Hog cholera		80–90	"
Rickettsiae			
Heart-water/veldt sickness (sheep and goats)	*Rickettsia ruminantium*	50–60	Aerosol; tick vectors

NAME OF DISEASE	CAUSATIVE AGENT	DEATH RATE (%) IN UNTREATED CASES OF NATURAL DISEASE	LIKELY MODE OF DISSEMINATION
Fungi			
Aspergillosis (poultry)	*Aspergillus fumigatus*	50–90	Dust/food contamination
Lumpy jaw (cattle)	*Actinomyces bovis*	50–90	Food contamination

ANTIPLANT AGENTS

NAME OF DISEASE	CAUSATIVE AGENT	DEATH RATE (%) IN UNTREATED CASES OF NATURAL DISEASE	LIKELY MODE OF DISSEMINATION
Viruses			
Tobacco mosaic			Aerosol
Sugar-beet curly top			Arthropod vectors
Corn stunt			
Hoja blanca (rice)			
Fiji disease (sugarcane)			Arthropod vectors
Potato yellow disease			
Bacteria			
Rice blight	*Xanthomonas oryzae*		
Corn blight	*Pseudomonas alboprecipitans*		
Sugarcane wilt	*Xanthomonas vasculorum*		Aerosol
Fungi			
Late blight of potato	*Phytophythora infestans*		Aerosol/dust
Coffee rust	*Hemileia vastatrix*		"
Maize rust	*Puccinia polysora*		"
Powdery mildew of cereals	*Erysiphe graminis*		"
Black stem rust of cereals†	*Puccinia graminis*	3–90	"

NAME OF DISEASE	CAUSATIVE AGENT	DEATH RATE (%) IN UNTREATED CASES OF NATURAL DISEASE	LIKELY MODE OF DISSEMINATION
Rice brown-spot disease	*Helmintho-sporium oryzae*		"
Rice Blast†	*Pyricularia oryzae*	50–90	"
Stripe rust of cereals	*Puccinia glumarum*		"

* Most or all of these have been studied intensively by the United States and possibly other nations. Many are still undergoing intensive study.

† Known to have been standardized and/or manufactured as BW agents in the past. No nation is verifiably known to stockpile BW agents currently.

Sources: SIPRI, *CB Weapons Today;* Geissler, *Biological and Toxin Weapons Today; Taber's Cyclopedic Medical Dictionary.*

TABLE 2A

CHARACTERISTICS OF SELECTED ANTIPERSONNEL CHEMICAL AGENTS

	COMMON/CODE NAME	CHEMICAL NAME	EFFECTS	LETHAL DOSE*
Harassing agents	CN CS†	2-chloraceto-phenone 2-chlorobenzyli-dine	Burning feeling, tears, respiratory difficulty, nausea	11,000
Incapacitating agents	BZ†	3-quinuclidinyl benzilate	Hallucina-tions, dis-orientation, giddiness	High
Incapacitating or lethal agents	chlorine		Respiratory difficulty, broncho-pneumonia	19,000
	mustard gas†	bis (2-chloroethyl) sulfide	Skin/eye blisters, lung damage, broncho-pneumonia	1,600

COMMON/CODE NAME	CHEMICAL NAME	EFFECTS	LETHAL DOSE*
tabun nerve gas† (GA)	ethyl NN-dimethyl-phosphoramide cyanidate	Sweating, vomiting, cramps, chest tightness, convulsions, coma, asphyxiation	400
sarin nerve gas† (GB)	iso-propyl-methyl-phosphoro-fluoridate	(Same)	100
soman nerve gas† (GD)	1,2,2-trimethyl-propyl methyl-phosphoro-fluoridate	(Same)	50
VX nerve gas†	ethyl S-2-diisopropyl-aminoethyl-methyl-phos-phorothiolate	(Same)	10

† Currently known to be stockpiled by some nations.
* Dosage is based on the number of milligrams/minute/cubic meter of airborne agent likely to kill 50 percent of unprotected people.

Sources: Murphy, et al., *No Fire No Thunder*; SIPRI, *CB Weapons Today*.

TABLE 3A

TOXINS MOST LIKELY TO BE USED AS WEAPONS*

TOXIN	SOURCE
Abrin	Plant
Aflatoxin	Fungus
Algal toxin with low molecular weight	Algae
Batrachotoxin	Frog
Botulin†	Bacterium
Coral toxins with low molecular weight	Corals
Ricin†	Plant
Saxitoxin†	Shellfish

TOXIN	SOURCE
Sea wasp toxin	Jellyfish
Staphylococcus enterotoxin	Bacteria
Tetanus toxin	Bacterium
Trichothecene mycotoxins	Fungus

* All are currently being studied as potential threats by the United States and probably other nations.
† Known to have at one time been standardized or manufactured as TW agents by the United States.

Sources: Geissler, *Biological and Toxin Weapons Today*; SIPRI, *CB Weapons Today.*

TABLE 4A

RELATIVE LETHALITY OF SELECTED CHEMICALS AND NATURALLY OCCURRING POISONS*

CHEMICALS	TOXIN—SOURCE ORGANISM
	Shiga toxin—*Shigella dysenteriae* (bacteria)
	Tetanus toxin—*Clostridium tetani* (bacteria)
	Botulinum toxin—*Clostridium botulinum* (bacteria)
	Ricin—castor beans
Dioxin†	
	Tetrodotoxin—puffer fish
VX (nerve gas)	
Soman (nerve gas)	
Sarin (nerve gas)	
	Aflatoxin—*Aspergillus* (fungus)
Tabun (nerve gas)	
	T2 toxin—Trichothecene (fungus)
Mustard gas	snake venom toxin—rattlesnake
Phosgene	
Chlorine	Bee venom toxin—honeybee

* The substances highest on the chart are the most lethal, or those fatal in the smallest doses. Therefore, Shiga toxin is the most lethal. The placement is nominal. For example, tetanus toxin is nearly 5 times as toxic, by weight, as botulinum toxin, 300 times as toxic as dioxin, and 1.5 million times as toxic as tabun.
† Dioxin is a contaminant of some herbicides, such as Agent Orange.

Sources: SIPRI, *CB Weapons Today,* Murphy, et al., *No Fire No Thunder.*

APPENDIX 3

Protocol for the Prohibition of the Use in War of Asphyxiating, Poisonous or Other Gases, and of Bacteriological Methods of Warfare

Signed at Geneva June 17, 1925
Entered into force February 8, 1928
Ratification by the U.S. Senate December 16, 1974
Ratified by U.S. President January 22, 1975
U.S. Ratification deposited with the Government of France April 10, 1975
Proclaimed by U.S. President April 29, 1975

The Undersigned Plenipotentiaries, in the name of their respective Governments:

Whereas the use in war of asphyxiating, poisonous or other gases, and of all analogous liquids, materials or devices, has been justly condemned by the general opinion of the civilized world; and

Whereas the prohibition of such use has been declared in Treaties to which the majority of Powers of the World are Parties; and

To the end that this prohibition shall be universally accepted as part of International Law, binding alike the conscience and the practice of nations;

Declare:

That the High Contracting Parties, so far as they are not already Parties to Treaties prohibiting such use, accept this prohibition, agree to extend this prohibition to the use of bacteriological methods of warfare and agree to be bound as between themselves according to the terms of this declaration.

The High Contracting Parties will exert every effort to induce other States to accede to the present Protocol. Such accession will be notified to the Government of the French Republic, and by the latter to all signatory and acceding Powers, and will take effect on the date of the notification by the Government of the French Republic.

The present Protocol, of which the French and English texts are both authentic, shall be ratified as soon as possible. It shall bear today's date.

The ratifications of the present Protocol shall be deposited to the Government of the French Republic, which will at once notify the deposit of such ratification to each of the signatory and acceding Powers.

The instruments of ratification of and accession to the present Protocol will remain deposited in the archives of the Government of the French Republic.

The present Protocol will come into force for each signatory Power as from the date of deposit of its ratification, and, from that moment, each Power will be bound as regards other powers which have already deposited their ratifications.

IN WITNESS WHEREOF the Plenipotentiaries have signed the present Protocol.

DONE at Geneva in a single copy, this seventeenth day of June, One Thousand Nine Hundred and Twenty-Five.

Convention on the Prohibition of the Development, Production and Stockpiling of Bacteriological (Biological) and Toxin Weapons and on Their Destruction

Signed at Washington and Moscow April 10, 1972
Ratification advised by U.S. Senate December 16, 1974
Ratified by U.S. President January 22, 1975
U.S. ratification deposited at Washington, London, and Moscow
March 26, 1975
Proclaimed by U.S. President March 26, 1975
Entered into force March 26, 1975

The States Parties to this Convention,

Determined to act with a view to achieving effective progress towards general and complete disarmament, including the prohibition of the development, production and stockpiling of chemical and bacteriological (biological) weapons and their elimination, through effective measures, will facilitate the achievement of general and complete disarmament under strict and effective international control.

Recognizing the important significance of the Protocol for the Prohibition of the Use in War of Asphyxiating, Poisonous or Other Gases, and of Bacteriological Methods of Warfare, signed at Geneva on June 17, 1925, and conscious also of the contribution which the said Protocol has already made, and continues to make, to mitigating the horrors of war,

Reaffirming their adherence to the principles and objectives of that Protocol and calling upon all States to comply strictly with them,

Recalling that the General Assembly of the United Nations has repeatedly condemned all actions contrary to the principles and objectives of the Geneva Protocol of June 17, 1925,

Desiring to contribute to the strengthening of confidence between peoples and the general improvement of the international atmosphere,

Desiring also to contribute to the realization of the purposes and principles of the Charter of the United Nations,

Convinced of the importance and urgency of eliminating from the arsenals of States, through effective measures, such dangerous weapons of mass destruction as those using chemical or bacteriological (biological) agents,

Recognizing that an agreement on the prohibition of bacteriological (biological) and toxin weapons represents a first possible step towards the achievement of agreement on effective measures also for the prohibition of the development, production and stockpiling of chemical weapons, and determined to continue negotiations to that end,

Determined, for the sake of all mankind, to exclude completely the possibility of bacteriological (biological) agents and toxins being used as weapons,

Convinced that such use would be repugnant to the conscience of mankind and that no effort should be spared to minimize this risk,

Having agreed as follows:

Article I

Each State Party to this Convention undertakes never in any circumstances to develop, produce, stockpile or otherwise acquire or retain:

(1) Microbial or other biological agents, or toxins whatever their origin or method of production, of types and in quantities that have no justification for prophylactic, protective or other peaceful purposes;

(2) Weapons, equipment or means of delivery designed to use such agents or toxins for hostile purposes or in armed conflict.

Article II

Each State Party to this Convention undertakes to destroy, or to divert to peaceful purposes, as soon as possible but not later than nine months after the entry into force of the Convention, all agents, toxins, weapons, equipment and means of delivery specified in article I of the Convention, which are in its possession or under its jurisdiction or control. In implementing the provisions of this article all necessary safety precautions shall be observed to protect populations and the environment.

Article III

Each State Party to this Convention undertakes not to transfer to any recipient whatsoever, directly or indirectly, and not in any way to assist, encourage, or induce any State, group of States or international organizations to manufacture or otherwise acquire any of the agents, toxins, weapons, equipment or means of delivery specified in article I of the Convention.

Article IV

Each State Party to this Convention shall in accordance with its constitutional processes, take any necessary measures to prohibit and prevent the development, production, stockpiling, acquisition, or retention of the agents, toxins, weapons, equipment and means of delivery specified in article I of the Convention, within the territory of such State, under its jurisdiction or under its control anywhere.

Article V

The States Parties to this Convention undertake to consult one another and to cooperate in solving any problems which may arise in relation to the objective of, or in the application of the provisions of, the Convention. Consultation and cooperation pursuant to this article may also be undertaken through appropriate international procedures within the framework of the United Nations and in accordance with its Charter.

Article VI

(1) Any State Party to this Convention which finds that any other State Party is acting in breach of obligations deriving from the provisions of the Convention may lodge a complaint with the Security Council of the United Nations. Such a complaint should include all possible evidence confirming its validity, as well as a request for its consideration by the Security Council.

(2) Each State Party to this Convention undertakes to cooperate in carrying out any investigation which the Security Council may initiate, in accordance with the provisions of the Charter of the United Nations, on the basis of the complaint received by the Council. The Security Council shall inform the States Parties to the Convention of the results of the investigation.

Article VII

Each State Party to this Convention undertakes to provide or support assistance, in accordance with the United Nations Charter, to any Party to the Convention which so requests, if the Security Council decides that such Party has been exposed to danger as a result of violation of the Convention.

Article VIII

Nothing in this Convention shall be interpreted in any way limiting or detracting from the obligations assumed by any State under the Protocol for the Prohibition of the Uses in War of Asphyxiating, Poisonous or other Gases, and of Bacteriological Methods of Warfare, signed at Geneva on June 17, 1925.

Article IX

Each State Party to this Convention affirms the recognized objective of effective prohibition of chemical weapons and, to this end, undertakes to continue negotiations in good faith with a view to reaching early agreement on effective measures for the prohibition of their development, production and stockpiling and for their destruction, and on appropriate measures concerning equipment and means of delivery specifically designed for the production or use of chemical agents for weapons purposes.

Article X

(1) The States Parties to this Convention undertake to facilitate, and have the right to participate in, the fullest possible exchange of equipment, materials and scientific and technological information for the use of bacteriological (biological) agents and toxins for peaceful purposes. Parties to the Convention in a position to do so shall also cooperate in contributing individually or together with other States or international organizations to the further development and applications of scientific discoveries in the field of bacteriology (biology) for prevention of disease, or for other peaceful purposes.

(2) This Convention shall be implemented in a matter designed to avoid hampering the economic or technological development of States Parties to the Convention or international cooperation in the field of peaceful bacteriological (biological) activities, including the international exchange of bacteriological (biological) agents and toxins and equipment for the processing, use or production of bacteriological (biological) agents and toxins for peaceful purposes in accordance with the provisions of the Convention.

Article XI

Any State Party may propose amendments to this Convention. Amendments shall enter into force for each State Party accepting the amendments upon their acceptance by a majority of the States Parties to the Convention and thereafter for each remaining State Party on the date of acceptance by it.

Article XII

Five years after the entry into force of this Convention, or earlier if it is requested by a majority of Parties to the Convention by submitting a proposal to this effect to the Depositary Governments, a conference of States Parties to the Convention shall be held at Geneva, Switzerland, to review the operation of the Convention, with a view to assuring that the purposes of the preamble and the provisions of the Convention, including the provisions concerning negotiations on chemical weapons, are being realized. Such review shall take into account any new scientific and technological developments relevant to the Convention.

Article XIII

(1) This Convention shall be of unlimited duration.

(2) Each State Party to this Convention shall in exercising its national sovereignty have the right to withdraw from the Convention if it decides that extraordinary events, related to the subject matter of the Convention, have jeopardized the supreme interests of its country. It shall give notice of such withdrawal to all other State Parties to the Convention and to the United Nations Security Council three months in advance. Such notice shall include a statement of the extraordinary events it regards as having jeopardized its supreme interests.

Article XIV

(1) This Convention shall be open to all States for signature. Any State which does not sign the Convention before its entry into force in accordance with paragraph (3) of this Article may accede to it at any time.

(2) This Convention shall be subject to ratification by signatory States. Instruments of ratification and instruments of accession shall be deposited with the Governments of the United States of America, the United Kingdom of Great Britain and Northern Ireland and the Union of Soviet Socialist Republics, which are hereby designated the Depositary Governments.

(3) This Convention shall enter into force after the deposit of instruments of ratification by twenty-two Governments, including the Governments designated as Depositaries of the Convention.

(4) For States whose instruments of ratification or accession are deposited subsequent to the entry into force of this Convention, it shall enter into force on the date of the deposit of their instruments of ratification or accession.

(5) The Depositary Governments shall promptly inform all signatory and acceding States of the date of each signature, the date of deposit of each instrument of ratification or of accession and the date of the entry into force of this Convention, and of the receipt of other notices.

(6) This Convention shall be registered by the Depositary Governments pursuant to Article 102 of the Charter of the United Nations.

Article XV

This Convention, the English, Russian, French, Spanish and Chinese texts of which are equally authentic, shall be deposited in the archives of the Depositary Governments. Duly certified copies of the Convention shall be transmitted by the Depositary Governments to the Governments of the Signatory and acceding states.

IN WITNESS WHEREOF the undersigned, duly authorized, have signed this convention.

DONE in triplicate, at the cities of Washington, London and Moscow, this tenth day of April, one thousand nine hundred and seventy-two.

Source: U.S. Arms Control and Disarmament Agency.

IMPLEMENTATION STATUS OF THE GENEVA PROTOCOL AND BW CONVENTION

(As of January 1, 1986)

Number of parties:
Geneva Protocol 108
BW Convention 103

Note: The below table records years of ratification, accession or succession. *S* refers to nations that signed the agreements but have not taken further action. The footnotes paraphrase original qualifications of the parties, but the wording is close to the original.

STATE	GENEVA PROTOCOL	BW CONVENTION
Afghanistan		**1975**
Algeria		
Antigua and Barbuda		
Argentina	**1969**	**1979**
Australia	**1930**[1]	**1977**
Austria	**1928**	**1973**[1]
Bahamas		
Bangladesh		**1985**
Barbados	**1976**[2]	**1973**
Belgium	**1928**[1]	**1979**

STATE	GENEVA PROTOCOL	BW CONVENTION
Belize		
Benin		1975
Bhutan	1978	1978
Bolivia	1985	1975
Botswana		S
Brazil	1970	1973
Brunei Darussalam		
Bulgaria	1934[1]	1972
Burkina Faso (formerly Upper Volta)	1971	
Burma		S
Burundi		S
Byelorussia	1970[3]	1975
Cameroon		
Canada	1930[1]	1972
Cape Verde		1977
Central African Republic	1970	S
Chad		
Chile	1935[1]	1980
China	1929[4]	1984[2]
Colombia		1983
Congo		1978
Costa Rica		1973
Cuba	1966	1976
Cyprus	1966[2]	1973
Czechoslovakia	1938[5]	1973
Denmark	1930	1973
Dominica		
Dominican Republic	1970	1973
Ecuador	1970	1975
Egypt	1928	S
El Salvador	S	S
Equatorial Guinea		
Ethiopia	1935	1975
Fiji	1973[1,2]	1973
Finland	1929	1974
France	1926[1]	1984
Gabon		S
Gambia	1966[2]	S
German Democratic Republic	1929	1972
Germany, Federal Republic of	1929	1983[3]

STATE	GENEVA PROTOCOL	BW CONVENTION
Ghana	1967	1975
Greece	1931	1975
Grenada		
Guatemala	1983	1973
Guinea		
Guinea-Bissau		1976
Guyana		S
Haiti		S
Holy See (Vatican City)	1966	
Honduras		1979
Hungary	1952	1972
Iceland	1967	1973
India	1930[1]	1974[4]
Indonesia	1971[2]	S
Iran	1929	1973
Iraq	1931[1]	S
Ireland	1930[6]	1972[5]
Israel	1969[7]	
Italy	1928	1975
Ivory Coast	1970	S
Jamaica	1970[2]	1975
Japan	1970	1982
Jordan	1977[8]	1975
Kampuchea	1983[9]	1983
Kenya	1970	1976
Kiribati		
Korea, Democratic People's Republic		
Korea, Republic of		S[6]
Kuwait	1971[10]	1972[7]
Lao People's Democratic Republic		1973
Lebanon	1969	1975
Lesotho	1972[2]	1977
Liberia	1927	S
Libya	1971[11]	1982
Liechtenstein		
Luxembourg	1936	1976
Madagascar	1967	S
Malawi	1970	S
Malaysia	1970	S

STATE	GENEVA PROTOCOL	BW CONVENTION
Maldives	1966[2]	
Mali		S
Malta	1970[2]	1975
Mauritania		
Mauritius	1970[2]	1972
Mexico	1932	1974[8]
Monaco	1967	
Mongolia	1968[12]	1972
Morocco	1970	S
Nauru		
Nepal	1969	S
Netherlands	1930[13]	1981
New Zealand	1930[1]	1972
Nicaragua	S	1975
Niger	1967[2]	1972
Nigeria	1968[1]	1973
Norway	1932	1973
Pakistan	1960[2]	1974
Panama	1970	1974
Papua New Guinea	1981[1]	1980
Paraguay	1933[14]	1976
Peru	1985	1985
Philippines	1973	1973
Poland	1929	1973
Portugal	1930[1]	1975
Qatar	1976	1975
Romania	1921[1]	1979
Rwanda	1964[2]	1975
Saint Lucia		
Saint Vincent and the Grenadines		
Samoa		
San Marino		1975
São Tomé and Principe		1979
Saudi Arabia	1971	1972
Senegal	1977	1975
Seychelles		1979
Sierra Leone	1967	1976
Singapore		1975
Solomon Islands		1981[9]
Somalia		S
South Africa	1930[1]	1975

STATE	GENEVA PROTOCOL	BW CONVENTION
Spain	1929[15]	1979
Sri Lanka	1954	S
Sudan	1980	
Suriname		
Swaziland		
Sweden	1930	1976
Switzerland	1932	1976[10]
Syria	1968[16]	S
Taiwan		1973[11]
Tanzania	1963[17]	S
Thailand	1931	1975
Togo	1971	1976
Tonga	1971	1976
Trinidad and Tobago	1970[2]	
Tunisia	1967	1973
Turkey	1929	1974
Tuvalu		
Uganda	1965	
Ukraine		1975
Union of Soviet Socialist Republics	1928[18]	1975
United Arab Emirates		S
United Kingdom	1930[1]	1975[12]
United States of America	1975[19]	1975
Uruguay	1977	1981
Venezuela	1928	1978
Vietnam	1980[1]	1980
Yemen Arab Republic	1971	S
Yemen, People's Democratic Republic of		1979
Yugoslavia	1929[20]	1973
Zaire		1977
Zambia		

Footnotes to Geneva Protocol

1 The protocol is binding on this state only as regards states that have signed and ratified or acceded to it. The protocol will cease to be binding on this state in regard to any enemy state whose armed forces or whose allies fail to respect the prohibitions laid down in the protocol.

2 Notification of succession. (In notifying its succession to the obligations contracted in 1930 by the United Kingdom, Barbados stated that as far as it was concerned, the reservation made by the U.K. was to be considered withdrawn.)

3 In a note of March 2, 1970, submitted at the United Nations, Byelorussia stated that "it recognizes itself to be a party" to the protocol.

4 On July 13, 1952, the People's Republic of China issued a statement recognizing as binding upon it the accession to the protocol in the name of China. China considers itself bound by the protocol on condition of reciprocity on the part of all the other contracting and acceding powers.

5 Czechoslovakia shall cease to be bound by this protocol toward any state whose armed forces, or the armed forces of whose allies, fail to respect the prohibitions laid down in the protocol.

6 The government of Ireland does not intend to assume, by this accession, any obligation except toward the states having signed and ratified this protocol or which have finally acceded thereto, and should the armed forces or the allies of an enemy state fail to respect the protocol, the government of Ireland would cease to be bound by the said protocol in regard to such state. In February 1972 Ireland declared that it had decided to withdraw the above reservations made at the time of accession to the protocol.

7 The protocol is binding on Israel only as regards states which have signed and ratified or acceded to it. The protocol shall cease to be binding on Israel as regards any enemy state whose armed forces, or the armed forces of whose allies, or the regular or irregular forces, or groups or individuals operating from its territory, fail to respect the prohibitions that are the object of the protocol.

8 The accession by Jordan to the protocol does not in any way imply recognition of Israel. Jordan undertakes to respect the obligations contained in the protocol with regard to states that have undertaken similar commitments. It is not bound by the protocol as regards states whose armed forces, regular or irregular, do not respect the provisions of the protocol.

9 The accession was made on behalf of the coalition government of Democratic Kampuchea (the government-in-exile), with a statement that the protocol will cease to be binding on it in regard to any enemy state whose armed forces or whose allies fail to respect the prohibitions laid down in the protocol. The French government declared that as a party to the Geneva Protocol (but not as the depositary) it considers this accession to have no effect. A similar statement was made by the governments of Australia, Bulgaria, Cuba, Czechoslovakia, the German Democratic Republic, Hungary, Mauritius, Netherlands, Poland, Romania, the Union of Soviet Socialist Republics, and Vietnam, which do not recognize the coalition government of Kampuchea.

10 The accession of Kuwait to the protocol does not in any way imply recognition of Israel or the establishment of relations with the latter on the basis of the present protocol. In case of breach of the prohibition laid down in this protocol by any of the parties, Kuwait will not be bound, with regard to the party committing the breach, to apply the provisions of this protocol.

11 The accession to the protocol does not imply recognition of Israel. The protocol is binding on Libya only as regards states that are effectively bound by it and will cease to be binding on Libya as regards states whose armed forces, or the armed forces of whose allies, fail to respect the prohibitions that are the object of this protocol.

12 In the case of violation of this prohibition by any state in relation to Mongolia or

its allies, the government of Mongolia shall not consider itself bound by the obligations of the protocol toward that state.

13 As regards the use in war of asphyxiating, poisonous, or other gases and all analogous liquids, materials, or devices, this protocol shall cease to be binding on the Netherlands with regard to any enemy state whose armed forces or whose allies fail to respect the prohibitions laid down in the protocol.

14 This is the date of receipt of Paraguay's instrument of accession. The date of notification by the depositary government "for the purpose of regularization" is 1969.

15 Spain declared the protocol as binding ipso facto, without special agreement with respect to any other member of state accepting and observing the same obligation—that is, on condition of reciprocity.

16 The accession by Syria to the protocol does not in any case imply recognition of Israel or lead to the establishment of relations with the latter concerning the provisions laid down in the protocol.

17 The protocol, signed in 1929 in the name of China, is valid for Taiwan, which is part of China.

18 The protocol binds the Union of Soviet Socialist Republics in relation to the states that have signed and ratified or that have definitely acceded to the protocol. The protocol shall cease to be binding on the USSR in regard to any enemy state whose armed forces or whose allies de jure or in fact do not respect the prohibitions that are the object of this protocol.

19 The protocol shall cease to be binding on the United States of America with respect to the use in war of asphyxiating, poisonous, or other gases, and of all analogous liquids, materials, or devices, in regard to an enemy state if such state or any of its allies fail to respect the prohibitions laid down in the protocol.

20 The protocol shall cease to be binding on Yugoslavia in regard to any enemy state whose armed forces or whose allies fail to respect the prohibitions that are the object of the Protocol.

Footnotes to the BW Convention

1 Considering the obligations resulting from its status as a permanently neutral state, Austria declares a reservation to the effect that its cooperation within the framework of this convention cannot exceed the limits determined by the status of permanent neutrality and membership with the United Nations.

2 China stated that the BW Convention has the following defects: It fails explicitly to prohibit the use of biological weapons; it does not provide for "concrete and effective" measures of supervision and verification; and it lacks measures of sanctions in case of violation of the convention. The Chinese government hopes that these defects will be corrected at an appropriate time and also that a convention for complete prohibition of chemical weapons will soon be concluded. The signature and ratification of the convention by the Taiwan authorities in the name of China are considered illegal and null and void.

3 On depositing its instrument of ratification, the Federal Republic of Germany stated that a major shortcoming of the BW Convention is that it does not contain any provisions for verifying compliance with its essential obligations. The federal government considers the right to lodge a complaint with the UN Security Council to be an

inadequate arrangement. It would welcome the establishment of an independent international committee of experts able to carry out impartial investigations when doubts arise whether the convention is being complied with.

4 In a statement made on the occasion of the signature of the convention, India reiterated its understanding that the objective of the convention is to eliminate biological and toxin weapons, thereby excluding completely the possibility of their use, and that the exemption with regard to biological agents or toxins, which would be permitted for prophylactic, protective, or other peaceful purposes, would not in any way create a loophole in regard to the production or retention of biological and toxin weapons. Also, any assistance which might be furnished under the terms of the convention would be of a medical or humanitarian nature and in conformity with the UN Charter. The statement was repeated at the time of deposit of the instrument of ratification.

5 Ireland considers that the convention would be undermined if the reservations made by parties to the 1925 Geneva Protocol were allowed to stand, as the prohibition of possession is incompatible with the right to retaliate, and that there should be an absolute and universal prohibition of the use of the weapons in question. Ireland notified the depositary government for the Geneva Protocol of the withdrawal of its reservations to the protocol, made at the time of accession in 1930. The withdrawal applies to chemical as well as to bacteriological (biological) and toxin agents of warfare.

6 The Republic of Korea stated that the signing of the convention does not in any way mean or imply recognition of any territory or regime that has not been recognized by the Republic of Korea as a state or government.

7 In the understanding of Kuwait, its ratification of the convention does not in any way imply its recognition of Israel, nor does it oblige it to apply the provisions of the conventions in respect to the said country.

8 Mexico considers that the convention is only a first step toward an agreement prohibiting also the development, production, and stockpiling of all chemical weapons and notes the fact that the convention contains an express commitment to continue negotiations in good faith with the aim of arriving at such an agreement.

9 Notification of succession.

10 The ratification of Switzerland contains the following reservations: (1) Owing to the fact that the convention also applies to weapons, equipment, or means of delivery designed to use biological agents or toxins, the delimitation of its scope of application can cause difficulties since there are scarcely any weapons, equipment, or means of delivery peculiar to such use; therefore, Switzerland reserves the right to decide for itself what auxiliary means fall within that definition. (2) By reason of the obligations resulting from its status as a perpetually neutral state, Switzerland is bound to make the general reservation that its collaboration within the framework of this convention cannot go beyond the terms prescribed by that status. This reservation refers especially to Article VII of the convention as well as to any similar clause that could replace or supplement that provision of the convention.

In a note of August 18, 1976, addressed to the Swiss ambassador, the U.S. secretary of state stated the following view of the U.S. government with regard to the first reservation: The prohibition would apply only to (a) weapons, equipment, and means of delivery the design of which indicated that they could have no other use than that specified, and (b) weapons, equipment, and means of delivery the design of which indicated that they were specifically intended to be capable of the use specified. The government of the United States shares the view of the government of Switzerland that there are few weapons, equipment, or means of delivery peculiar to the uses referred to. It does not, however, believe that it would be appropriate, on this ground alone, for states to reserve unilaterally the right to decide which weapons, equipment, or means of delivery fell within the definition. Therefore, while acknowledging the entry into force of the convention between itself and the government of Switzerland, the U.S. government enters its objection to this reservation.

11 The deposit of the instrument of ratification by Taiwan is considered by the Soviet Union an illegal act because the government of the People's Republic of China is regarded by the Soviet Union as the sole representative of China.

12 The United Kingdom recalled its view that if a regime is not recognized as the government of a state, neither signature nor deposit of any instrument by it nor notification of any of those acts will bring about recognition of that regime by any other state.
Source: Stockholm International Peace Research Institute.

SOURCE NOTES

(Note: Magazine page numbers refer to first page of article. For declassified government documents, date of declassification is given if available.)

Chapter 1

Pages 16: Feith comments are from: "Testimony on Biological and Toxin Weapons Before the Subcommittee on Oversight and Evaluation of the House Permanent Select Committee on Intelligence," August 8, 1986.

Pages 16–17: A good description of the Cohen-Boyer collaboration can be found in: *Time* (March 9, 1981), p. 50. One of the many good sources describing recombinant DNA technology and its potential is: Congressional Office of Technology Assessment, *Commercial Biotechnology: An International Analysis* (1984), pp. 33–57.

Pages 19–20: For a comprehensive treatment of the research moratorium, see: Sheldon Krimsky, *Genetic Alchemy* (1982), pp. 58–96. For details of containment rules, see: Department of Health and Human Services, "Guidelines for Research Involving Recombinant DNA Molecules" (1986). For a description of the experiment involving giant mice, see: *Nature* (December 16, 1982), p. 611.

Pages 20–22: Definitions of biological, chemical, and toxin weapons are drawn from: Stockholm International Peace Research Institute (SIPRI), *World Arma-*

ments and Disarmament, SIPRI 1984 Yearbook, p. 422; and Erhard Geissler, ed., *Biological and Toxin Weapons Today* (1986), pp. 4–7.

Pages 22–25: Overall applications of biotechnology are drawn from Geissler, op. cit., pp. 27–32; SIPRI, op. cit., pp. 426–33. Production improvements and "biochemical" weapons are discussed in Department of Defense (DOD), "Biological Defense Program: Report to the Committee of Appropriations, House of Representatives" (1986), pp. 8–9. For "ethnic weapons" background, see, for example: *Military Review* (November 1980), p. 3.

Pages 25–27: BW and CW spending figures are drawn from: DOD, "Annual Report on Chemical Warfare-Biological Defense Program Obligations" (fiscal years 1976–85). Life sciences spending figures are drawn from National Science Foundation, "Federal Funds for Research and Development" (fiscal years 1973–87). Regarding U.S. claims of Soviet use of biotechnology for weaponry, see, for example: DOD, "Soviet Military Power" (April 1984), p. 73; and DOD, "Soviet Biological Warfare Threat" (1986). The range of U.S. BW research is detailed in: DOD, "Annual Report[s] on Chemical Warfare-Biological. . . ."

Chapter 2

Pages 28–29: Caffa description is drawn from: *Journal of the American Medical Association* (April 4, 1966), p. 179. For background on the British and Roman examples, see: Sean Murphy, et al., *No Fire No Thunder* (1984), p. 27. Ancient India description is from: *Journal of Chemical Education* (May 1948), p. 268.

Pages 29–34: Much of the historical information is from: Robert Harris and Jeremy Paxman, *A Higher Form of Killing* (1982), ch. 1 and 2. Historical information, including reproduction of Churchill documents, drawn from: *American Heritage* (August/September 1985), p. 40. For Hormats quote, see: *Wall Street Journal,* April 24, 1985 (letter).

Pages 34–35: Main sources for Japanese program: Harris and Paxman, op. cit., ch. 4; *San Francisco Chronicle—This World* (magazine section), November 3, 1985, p. 7; and *Bulletin of the Atomic Scientists* (October 1981), pp. 44. Japanese denial of atrocities reported by: Agence France Press, "Government Denies World War II Tests on POW's," August 13, 1985.

Page 36: Source for amounts of agents needed for attacks: SIPRI, *The Problem of Chemical and Biological Warfare,* vol. V, *The Prevention of CBW* (1971), p. 208.

Pages 36–38: BW research and development information drawn from Harris and Paxman, op. cit., ch. 4; and *American Heritage* (August/September 1985), p. 40. U.S. chemical stockpile figure drawn from: SIPRI, *The Problem of Chemical and Biological Warfare* , vol. II, *CB Weapons Today* (1973), p. 194.

Pages 38–39: Project "Who, Me?" is described in National Defense Research Committee, Division 19, "Final Report on Who, Me?" December 19, 1944 (declassified, 1983). For Oppenheimer example, see: *Technology Review* (May/June, 1985), p. 14. Merck report is reprinted in: Senate, "Biological Testing Involving Human Subjects by the Department of Defense, 1977," pp. 64–71.

Pages 40–42: Information on the Japanese BW program is drawn largely from: *San Francisco Chronicle—This World,* November 3, 1985, p. 7; and *Bulletin of the Atomic Scientists* (October 1981), p. 44.

Pages 42–43: A good description of Operation Blue Skies appears in: *Science* (January 13, 1967), p. 174.

Pages 43–44: Information on specific estimates of Soviet CBW strength drawn from: SIPRI, *CB Weapons Today,* pp. 174–175. Hersh's source is quoted in: Harris and Paxman, op. cit., p. 146. Dollar figures for U.S. program are from: SIPRI, *CB Weapons Today,* pp. 204–05. Figures on academic and corporate participation in U.S. program are from: Senate, "Biological Testing," pp. 80–100.

Pages 44–45: Mosquito tests are described in: United States Army Chemical Corps, "Summary of Major Events and Problems, Fiscal Year 1959" (declassified), pp. 101–105. Operation Whitecoat information drawn from: Senate, "Biological Testing," pp. 177–88; and *Ramparts* (December 1969), p. 21. U.S. stockpile figures are from: SIPRI, *CB Weapons Today,* p. 234.

Pages 45–46: Primary source for all field tests: Senate, "Biological Testing."

Page 45: Meselson is quoted in: *Ramparts* (December 1969), p. 21. University of Utah information drawn from a wide range of internal university documents and contracts. Dugway information is from: *Deseret News* (Salt Lake City), February 19, 1979.

Page 45: Best source on South Pacific tests: *Washington Post Magazine* (May 12, 1985), p. 8. Simulant testing has received much press attention, for example: *Washington Monthly* (July/August 1985), p. 38. Primary sources for simulant tests are declassified documents, such as: U.S. Army Biological Laboratories, Fort Detrick, "Miscellaneous Publication 7, Study US65SP," July 1965. Brookshire statement is from correspondence dated July 11, 1985, regarding Piller's Freedom of Information Act request.

Pages 46–48: For first strike information, see: SIPRI, *CB Weapons Today,* pp. 194–97. Hornig quote is from: "Memorandum for the President, Subject: Use of Biological Weapons," December 8, 1966 (declassified, July 21, 1983).

Pages 48–49: OSS activities are described in: Harris and Paxman, op. cit., p. 202. Korean episode is described in ibid. pp. 162–63; John Marks, *The Search for the "Manchurian Candidate"* (1980), pp. 125–26; and *Bulletin of the Atomic Scientists* (October 1981), p. 44. Jackson comments are from: U.S. Department of State, Office of Intelligence Research, "Intelligence Report No. 5997.2: The Effect of the Bacteriological Warfare Campaign," November 7, 1952 (declassified, November 8, 1977).

Pages 49–51: The major source for facts and figures on MKULTRA is: Marks, op. cit.

Page 49: Subproject 146 is described in declassified CIA memorandums dated September 9, 1963, April 9, 1964, and May 19, 1964. Castro assassination efforts are discussed in a declassified CIA memorandum dated August 7, 1975.

Page 50: The best source on BZ is: *Mother Jones* (May 1982), p. 14.

Pages 51–52: CIA inspector general document is reprinted in: Senate, "Biomedical and Behavioral Research, 1975," pp. 882–83. Chemical Corps officer's quote is from *Ramparts* (December 1969), p. 21. Sheep kill is described well in: *Science* (December 27, 1968), p. 1460.

Pages 52–53: Early Dugway information is from: *Deseret News,* series of articles February 19–22, 1979; and unpublished 1979 response by Dugway officials to questions posed by Dale Van Atta, then a *Deseret News* reporter. Dugway 1986 incidents described in: *In These Times,* January 14–20, 1987. Rocky Mountain nerve gas disposal problems are drawn from: Harris and Paxman, op. cit., p. 217.

Pages 53–54: The official army count of BW research accidents is listed in: Senate, "Biological Testing," p. 152. Figures from the Fort Detrick safety officer are from: DOD, Fort Detrick, "Miscellaneous Publication 2: Microbiological Causal Factors in Laboratory Accidents and Infections," Table 7, April, 1965. Fort Greeley quote is from *Ramparts* (December 1969), p. 21. VEE episode is from: House of Representatives, "Environmental Dangers of Open-Air Testing of Chemicals" (1969), pp. 6–7, 51–55. Hodge quote is from *Deseret News,* February 21, 1979.

Chapter 3

Pages 56–57: Nixon statement is reproduced in: President's Chemical Warfare Review Commission, "Report of," June 1985, p. 91. Disarmament negotiator quoted in: SIPRI, *CB Weapons Today,* p. 157.

Pages 58–59: Information on Western and WTO CW capabilities is drawn from: *SIPRI Yearbook 1982,* pp. XXXIV, 323–34; *SIPRI Yearbook 1985,* p. 166–67; and *Bulletin of the Atomic Scientists* (November 1986), p. 28.

Page 59: Leath is quoted in: *New York Times,* September 16, 1983. Estimates of Soviet stockpile are reported in: *SIPRI Yearbook 1982,* p. XXXIV, 334–35; and

SIPRI Yearbook 1985, p. 170. CIA quote is from: *Boston Globe,* April 9, 1984. Hormats quote is from: *Christian Science Monitor,* April 24, 1986. For Kremlin statement on CW, see: *Science* (April 17, 1987), p. 252.

Pages 60–61: Information on the status of the U.S. chemical arsenal is from: *SIPRI Yearbook 1982,* pp. XXXIII, 324–28, 347–51; *SIPRI Yearbook 1985,* p. 170. Sources for Table 1 are: SIPRI, *CB Weapons Today,* pp. 204–05, 233; and DOD, "Annual Report[s] on Chemical Warfare-Biological . . ." (1976, 1982, 1984, 1985).

Pages 61–63: The binary nerve gas controversy was heavily reported. See, for example: *SIPRI Yearbook 1984,* pp. 321, 327; *Los Angeles Times,* June 20, July 28, 1985; *San Francisco Chronicle,* May 23 and August 14, 1986; *New York Times,* May 24, 1986; and the *Nation* (June 21, 1986), p. 841. Conference committee decisions are from: House of Representatives, "National Defense Authorization Act for Fiscal Year 1987, Conference Report" (No. 99-1001), October 14, 1986. Kerry quote is from: *Bulletin of the Atomic Scientists* (November 1986), p. 28.

Page 63: CW proliferation is discussed in: *Chemical and Engineering News* (April 14, 1986), p. 8; and General Accounting Office, "Chemical Warfare: Many Unanswered Questions," April 29, 1983, p. 88. For information on trade restrictions, see: *SIPRI Yearbook 1985,* pp. 171–74; and Nerve Center (Oakland, Calif.), *Unmask* (October 1985), p. 3.

Page 64: CIA toxin scandal background is from: Senate, "Unauthorized Storage of Toxic Agents," September 16–18, 1985; Karamessines quote appears on p. 189.

Pages 65–69: Table 2 is derived from dozens of U.S. and overseas press reports. For yellow rain data, we used a number of U.S. government publications, such as: Department of State, "Chemical Warfare in Southeast Asia and Afghanistan: An Update" (Special Report No. 104), November 1982. Also used extensively: *SIPRI Yearbooks 1982–1985.*

Pages 69–71: Vietnam data primarily drawn from: Senate, "The Geneva Protocol of 1925," March 5, 16, 18, 19, 22, 26, 1971; for Rogers quotes, see: p. 45; for Meselson quote, see: p. 363. Additional sources include: *Science for the People* (September–October 1983), p. 11; *Bulletin of the Atomic Scientists* (February 1981), p. 38. The effects of Agent Orange on GIs is described in: *Southeast Asia Chronicle* (June 1983), p. 24; and *New York Times,* August 29, 1986. For Ford policy, see: "Executive Order 11850: Renunciation of Certain Uses in War of Chemical Herbicides and Riot Control Agents," *Federal Register,* April 10, 1975.

Pages 71–72: Iraq information primarily drawn from: *SIPRI Yearbook 1985,* pp. 171 (Indian expert's quote), 181–83, appendix; and various press reports, such as *New York Times,* March 19, 27, 1984, March 15, 1986. Roberts quote is from: *Washington Quarterly* (Winter 1985), p. 157. Cuban information primarily drawn from: *Denver Post* (*Post-Newsday* syndicate), January 9, 1977; and *Granma* (Cuban government newspaper), September 23, 1984, October 27, 1985, June 1, 1986.

Pages 73–76: Primary sources for yellow rain are as follows:

Overall: U.S. Department of State, "Chemical Warfare in Southeast Asia and Afghanistan" (Special Reports No. 98 and 104), March 22, 1982, November 1982; *Chemical and Engineering News* (January 9, 1984), p. 8.

Karnow report: *Atlantic Monthly* (October 1985), p. 67.

Physical evidence: *Scientific American* (September 1985), p. 128; *SIPRI Yearbook 1985,* p. 187; *Atlantic Monthly* (October 1985), p. 67; *Science* (July 4, 1986), p. 18; and *Arms Control Today* (September 1986), p. 31.

Victim testimony: *Southeast Asia Chronicle* (June 1983), p. 3; *Atlantic Monthly* (October 1985), p. 67; *Scientific American* (September 1985), p. 128; *Foreign Policy* (Fall 1987), p. 100.

Intelligence: *SIPRI Yearbook 1986,* p. 164; *Arms Control Today* (September 1986), p. 31.

Pages 77–78: For Nicaragua claims see dispatch from: *Pacific News Service* (San Francisco), January 3, 1986. Regarding vulnerability of developing nations, see: SIPRI, *CB Weapons Today,* pp. 142–48. For details on Egypt, see: *Boston Globe,* June 6, 1976.

Page 79: "Halfway house" concept is discussed in: *Washington Quarterly* (Winter 1985), p. 155.

Chapter 4

Pages 81–83: Relevant portions of The Hague, Brussels, Versailles, Washington, and Geneva pacts reprinted in: SIPRI, *CBW and the Law,* vol. III, *The Problem of Chemical and Biological Warfare,* pp. 151–53. For a detailed legal analysis of U.S. consideration of these pacts, see: Senate, "The Geneva Protocol of 1925," pp. 72–113; regarding early Pentagon lobbying, same document, p. 262. American Chemical Society leader's quote is from: Harris and Paxman, op. cit., pp. 45–46. Regarding recent Pentagon lobbying, see, for example: *New York Times,* August 4, 1986.

Pages 85–86: For UN votes, see: SIPRI, *Arms Control; A Survey and Appraisal of Multilateral Agreements* (1978), pp. 214–15. SIPRI quote is from: *SIPRI Yearbook 1982,* p. 321. Meselson quote is from: *Scientific American* (May 1970), p. 15.

Page 87: Primary source for Table 3: *SIPRI Yearbook 1982,* pp. 336–37. For Portugal data: U.S. Department of State, "Comite Four-Portuguese Territories [sic]" (cable), November 1970 (declassified, March 17, 1982). Other sources covered in previous notes.

Page 88: Newall is quoted in: Harris and Paxman, op. cit., p. 83. British CS Policy is discussed in: SIPRI, *The Prevention of CBW,* pp. 45–46; and *Scientific American* (May 1970), p. 15.

Pages 89–90: Committee on Disarmament negotiations are detailed in *SIPRI Yearbooks.* Verification problems are described well in: *Issues in Science and Technology* (Spring 1986), p. 102. U.S.-USSR negotiating problems have been widely reported, for example: *New York Times,* April 27, August 26, 1986. Robinson quote is from: *SIPRI Yearbook 1982,* p. 345.

Pages 90–91: For a discussion of the Libya example, see: *San Francisco Examiner,* March 30, 1986. Vachon quote is from: *Survival* (March–April 1984), p. 79. Hopeful signs in CW negotiations are noted in: *New York Times,* April 30, October 5, 1987. Superpower hegemony is discussed in: *SIPRI Yearbooks;* and *Survival* article noted above. Regarding German initiatives, see: *Arms Control Today* (September 1986), p. 14.

Page 92: Meselson quote is from: WGBH Television (Boston) *Nova,* "The Mystery of Yellow Rain" (No. 1111), October 30, 1984, transcript p. 21. Miettinen quote is from a lecture, University of California at San Francisco, February 1, 1985. Bush's votes are described in: *New York Times,* August 8, 1986. McGrory quote is from: *San Jose Mercury News,* April 8, 1984.

Chapter 5

Pages 93–95: List of criteria are adapted from: Raymond A. Zilinskas, "Managing the International Consequences of Recombinant DNA Research" (1981), pp. 341–44; *SIPRI Yearbook 1984,* p. 426; *Journal of Immunology* (May 1947), pp. 31–32; and Geissler, op. cit., pp. 21–22. Myxoma example is drawn from: Zilinskas, op. cit., pp. 339, 342, 345.

Pages 96–102: Primary sources for this section, by application:

Increased pathogenicity: Geissler, op. cit., pp. 25–28.

Defeating immunity and diagnostic testing: Ibid, p. 29.

Controllability: Zilinskas, op. cit., pp. 356, 359.

Drug resistance: Ibid., pp. 349–51; and Geissler, op. cit. p. 30.

Increased environmental survivability: Geissler, op. cit., pp. 32, 41; and SIPRI, *CB Weapons Today,* p. 285.

Ethnic weapons: *Military Review* (November 1970), p. 3; U.S. Army Chemical Corps, "Special Report No. 160: Contamination of a Portion of the Naval Supply System," November 12, 1951 (declassified, 1981). For background on coccidioidomycosis, see, for example: *Journal of the American Medical Association* (January 10, 1959), p. 99; and *Wisconsin Journal of Medicine* (August 1959), p. 471. For Epstein-Barr virus information, see: *SIPRI Yearbook 1984,* p. 433.

Biochemical weapons: DOD, "Biological Defense Program," ch. 1, pp. 8–9.

Novel toxin development and manufacturing methods: *SIPRI Yearbook 1984,* p. 432.

Pages 102–4: de Jong quote is from: *Medical World News* (October 28, 1985), p. 62. For background on vaccinia virus, see: Geissler, op. cit., p. 62, 167; and *SIPRI Yearbook 1984,* p. 428. Regarding DOD research into CW "vaccines," see, for example: DOD, "Annual Report . . . FY 1982," pp. 11, 13, 25, 28. Figures on Fort Detrick infection rates are from: DOD, Fort Detrick, "Miscellaneous Publication 2: Microbiological Causal Factors in Laboratory Accidents and Infections," Table 7, April, 1965.

Pages 105–6: Bioprocess quote is from: DOD, "Biological Defense Program," ch. 1, p. 9. Ultrasensors are detailed in: DOD, "Annual Report[s] on Chemical Warfare-Biological . . ."; and also see: *Genetic Engineering News* (March/April 1983), p. 28. Nose quote is from: *New York Times,* November 19, 1985.

Page 107: Army aerosol quote is from: DOD, "Annual Report . . . FY 1979," army biological section, p. 9, and is discussed extensively in later obligation reports. The standby legislation quote is from: *Military Medicine* (February 1963), p. 145; see also: article on p. 135.

Page 108: Army meteorological quote is from: DOD, "Biological Defense Program," ch. 1, p. 1. Rvachev information, including Longini quote, is from: *Environmental Action* (June 1984), p. 1; Leonid A. Rvachev, "Experiment on Pandemic Process Modeling (Part I)," unpublished; and unpublished correspondence from Rvachev to U.S. scientists (supplied by Richard Asinov).

Pages 111–13: Novick comments are from: Interview by Piller, November 11, 1986. King quote is from: "The Fallacies of 'Defensive' Biological Weapons Programs," unpublished, September 20, 1985, pp. 6–7. Hoppensteadt comments are from: Interview by Piller, December 19, 1986. Callaham/Tsipis quote is from: *Bulletin of the Atomic Scientists* (November 1978), p. 11. Meselson quote is from: *National Journal* (July 19, 1986), p. 1775.

Pages 114–15: FEMA quote is from: "Civilian Biological and Chemical Defense Program" (policy statement, undated). Tigertt quote is from: *Military Medicine* (February 1963), p. 135. Feith comments throughout this chapter are from: "Testimony on Biological and Toxin Weapons . . . ," August 8, 1986. Defense Science Board conclusion is from a classified report quoted in: DOD, "Biological Defense Program," ch. 1, p. 6. Review commission quote is from: "Report of the Chemical Warfare Review Commission," June 1985, p. 71. Army quote is from: DOD, "Biological Defense Program," ch. 1, p. 7. SIPRI quote is from: *SIPRI Yearbook 1982,* pp. 328–29.

Pages 116–18: Beisel quote is from: *San Jose Mercury News—West,* April 14, 1984, p. 12. Operation Whitecoat quote is from: *Ramparts* (December 1969), p. 21. Table 6 is adapted from: Geissler, op. cit., p. 67. King and Strauss quote appears in: ibid., p. 68. SIPRI quote is from: SIPRI, *The Prevention of CBW,* p. 155; and see also: *SIPRI Yearbook 1984,* pp. 439–40.

Pages 118–19: Creasy is quoted in: SIPRI, *The Prevention of CBW,* p. 278. Army quote is from DOD, "Biological Defense Program," ch. 1, p. 7. Huxsoll is quoted in *San Francisco Examiner,* December 16, 1984. King and Strauss quotes are from: Geissler, op. cit., p. 68, 69. Figures on the potency of BW agents are derived from: *NATO Letter,* vol. 18, no. 7–8 (1970), p. 17.

Pages 120–24: Civil defense estimate is by: UN Secretary-General, "Chemical and Bacteriological (Biological) Weapons and the Effects of Their Possible Use" (UN Publication A/7575/Rev. 1), p. 83, 1969. Huxsoll interview was conducted November 19, 1986. SIPRI observations are from: SIPRI, *The CB Weapons Today,* p. 330–31.

Pages 124–26: Klare quotes are from: *The Progressive* (December 1983), p. 31. Meselson 1971 quotes are from: Senate, "The Geneva Protocol of 1925," pp. 363–65. Meselson 1986 quote is from: *Arms Control Today* (September 1986), p. 31.

Chapter 6

Pages 127–29: For a delineation of CBW responsibilities for each DOD agency, see: DOD, "Directive Number 5160.5," May 1, 1985. For further background, see: Geissler, op cit., pp. 58–61. VEE quote is from: *Science* (July 30, 1971), p. 405. For examples of other spin-offs, see: *Ordnance* (January–February 1966). Kingsbury quote is from: *San Jose Mercury News—West,* April 14, 1984, p. 12.

Pages 134–35: Information on projects conducted by Biotech Research Labs, Louisville University, and Walter Reed Institute of Research was drawn from DOD forms 1498, with the following agency accession numbers, respectively: DA300961 (March 1, 1983), DA0G6650 (August 5, 1983), DAOA8436 (October 10, 1982). Falkow study was published in: *Nature* (September 19, 1985), p. 262. Rifkin lawsuit was filed in U.S. District Court for the District of Columbia, Civil Action 86-2436, September 23, 1986.

Pages 137–38: Huxsoll, King comments are from: Interviews by Piller, November 19, 4, 1986, respectively. University of Washington information is from: DOD Form 1498, agency accession no. DA302383, March 29, 1985.

Pages 138–40: Army quotes/information are from: "Military Construction, Army, Reprogramming Request" (project 0817300), undated; Army, "Information Paper" (DAMA-PPM-T), November 8, 1984; and Army, "Environmental Assessment [for aerosol test lab]," January 31, 1985. Goldstein, Baltimore quotes are from: *Science* (December 7, 1984), p. 1176. Novick quote is from an unpublished article by him. Sasser, Weinberger letters (unpublished) are dated October 31 and November 2, 1984, respectively.

Pages 140–43: Meselson, Curtiss, Baltimore, Orton quotes are from: *Science* (December 7, 1984), p. 1176. Novick quotes are from: "Response to Environmental Assessment of Proposed Dugway Biowar Laboratory," unpublished. Weinberger letter (unpublished) is dated November 2, 1984. For background on Rifkin suit, see: U.S. District Court for the District of Columbia, Civil Action No. 84-3542. Army quote is from: Army, "Information Paper" (DAMA-PPM-T), November 8, 1984.

Pages 143–44: First Army quote is from: Army, "Environmental Assessment [for aerosol test lab]" January 31, 1985. Second quote is from: *Science* (May 17, 1985), p. 827. Green injunction (Civil Action No. 84-3542) is dated May 31, 1985. Report to House of Representatives is: DOD, "Biological Defense Program."

Pages 144–46: For a discussion of U.S./NATO war-fighting strategy, see: Geissler, op. cit., pp. 74–81; Mechtersheimer quotes are on pp. 74–75, 77. Army field manual quote is from: Headquarters, Army Operations, "Field Manual No. 100-5," August 20, 1982. For DOD rDNA policy, see: James P. Wade, Jr., acting undersecretary of defense, memorandum, "DOD Research Activities in the Field of Recombinant Deoxyribonucleic Acid (DNA)," April 1, 1981. Huxsoll quotes are from: Interview by Piller, November 19, 1986.

Page 148: For Marks quotes, see: Marks, op. cit., pp. 61–62.

Pages 149–51: Huxsoll, Orton, King comments are from: Interviews by Piller, November 19, 1986, March 24, 1983, November 4, 1986, respectively. Novick quote is from: Testimony before the Subcommittee on Oversight and Investigations of the House Committee on Energy and Commerce, December 18, 1985. Meselson quote is from: *San Jose Mercury News—West,* April 14, 1984, p. 12. Regarding DOD-corporate proprietary arrangements, see: AMRIID, "The Cooperative Antiviral Drug Development Program" (promotional brochure), undated. Birkner quotes are from: *New York Times,* May 28, 1984.

Pages 152–53: Glick quote is from: *Foreign Policy* (Winter 1984–1985), p. 58. DTIC quote is from: DOD, "Control of Unclassified Technical Data with Military or Space Application" (DOD 5230.25-PH), May 1985. For background on Poindexter policy, see: *New York Times,* March 18, 19, 1987. DOD 1986 quote is from: *New York Times,* November 13, 1986.

Page 154: Regarding "black budget," see: *Philadelphia Inquirer,* February 8, 1987, and *New York Times,* February 27, 1986.

Pages 159–61: Robertson, Hoskins, Top quotes are from: *Wall Street Journal,* September 17, 1986. Agent Orange background and quote are from: *Southeast Asia Chronicle* (June 1983), p. 24. Regarding corporate reluctance on CW contracts, see: *Hartford Courant,* July 9, 1985. GAO quote is from: GAO, "Chemical Warfare: Progress and Problems in Defensive Capability" (GAO/PEMD-86-11), July 1986, pp. 56–57. King quote is from: Interview by Piller, November 4, 1986.

Chapter 7

Pages 163–64: For text of Genocide Convention, see: Geissler, op. cit., pp. 132-134. For text and implementation status of Environmental Modification Convention, see: U.S. Arms Control and Disarmament Agency, *Arms Control and Disarmament Agreements* (1982), pp. 193–200. For a discussion of these treaties, see, for example: *BioScience* (November 1985), p. 627, and *SIPRI Yearbook 1984,* p. 436.

Pages 164–66: King quote is from: Interview by Piller, November 4, 1986. Falk quotes are from: Geissler, op. cit., pp. 115–16. Huxsoll quote is from: Interview by Piller, November 19, 1986. For a discussion of the status of toxins, see: Geissler, op cit., pp. 52–55, 124–26; *SIPRI Yearbook 1984,* pp. 431, 442–44; and *Bulletin of the Atomic Scientists* (November 1983), p. 20.

Pages 166–68: For remarks of conferees and Kaplan, and views of Green Party, see: *Science* (October 10, 1986), p. 143. For Hamm quote, see: *Contemporary Review* (March 1985), p. 127. For Pugwash quotes, see: Pugwash (Chicago), "Statement of the Pugwash Executive Committee, Prepared for the 1986 Review Conference of the Biological Weapons Convention," January 1986. For background on first review conference, see: Zilinskas, op. cit., pp. 378–81.

Pages 168–69: Boyle quote is from: Francis A. Boyle, "The Reagan Administration's Policies Toward the Biological Weapons Convention: Comments Prepared for the Convention of the Committee for Responsible Genetics," November 13, 1986. For information on domestic implementing legislation in countries other than the United States, see: *SIPRI Yearbook 1976,* pp. 398–99. For background on U.S. legislation efforts, see: Zilinskas, op. cit., p. 366. Recent U.S. effort refers to: HR807, 100th Congress.

Pages 169–71: Meselson quotes are from: Interview by Piller, November 5, 1986. Best sources on the Sverdlovsk incident are: *Science* (September 26, 1980), p. 1501, and *Bulletin of the Atomic Scientists* (June/July 1983), p. 24. Medvedev quotes are from: *New Scientist* (July 13, 1980). Soviet explanations at second review conference are discussed in: *Bulletin of the Atomic Scientists* (January/February 1987), p. 40, and *Science* (October 10, 1986), p. 143.

Pages 171–72: For DIA quote, see: DOD, "Soviet Biological Warfare . . . ,"

p. 8. Weinberger quote is from: Letter to Senator James Sasser (unpublished), November 2, 1984. For examples of Anderson columns, see: *San Francisco Chronicle,* December 1, 4, 1984, August 27, 1984, and March 1, 1985. Kucewicz series appeared in: *Wall Street Journal,* April 23, 25, 27, and May 1, 3, 8, 10, 18, 1984. For Dashiell quote, see: *Science* (June 15, 1984), p. 1215.

Pages 172–73: First Meselson quote is from: Interview by Piller, November 5, 1986. Second Meselson quote is from: *Arms Control Today* (September 1986), p. 31. For Soviet allegations, see, for example: *SIPRI Yearbook 1983,* pp. 399–400; *Washington Post,* July 18 and November 1, 1986; *San Francisco Chronicle—This World,* April 12, 1987, p. 8.

Pages 173–76: General review conference sources: *Bulletin of the Atomic Scientists* (January/February 1987), p. 40; *Science* (October 10, 1986), p. 143, and Review Conference, "Final Declaration" (released by United Nations Information Service), September 26, 1986.

Pages 173–74: SIPRI quote is from: *SIPRI Yearbook 1986,* p. 178. Feith quotes are from: "Testimony on Biological and Toxin Weapons . . . ," August 8, 1986. Rosenberg quote is from: *Bulletin of the Atomic Scientists* (January/February 1987), p. 40.

Pages 174–76: Ad hoc committee work is drawn from: Ad Hoc Meeting of Scientific and Technical Experts from States Parties to the BW Convention, "Draft Report/Revision 3," April 15, 1987. Falk comments are from: Geissler, op. cit., pp. 110, 112. Novick quote is from: Interview by Piller, November 11, 1986. First Sims quote is from: *Armament & Disarmament Information Unit Report* (September/October 1986), p. 1. Second Sims quote is from: Unpublished paper, October 13, 1986.

Page 176: Mikulak quote is from: *Armament & Disarmament Information Unit Report* (September/October, 1986), p. 1; while his accusations against the Soviets are noted in: *Genetic Engineering News* (July/August 1984), p. 22. Defense Science Board is quoted in: DOD, "Biological Defense Program," ch. 1, p. 6. Holmes quote is from: *Science* (October 10, 1986), p. 143. The 1969 DOD quote is from: SIPRI, *CB Weapons Today,* pp. 314.

Page 176: For Dugway-related quote, see: DOD, "Biological Defense Program," ch. 4, p. 2. SIPRI quote is from: *SIPRI Yearbook 1985,* p. 178.

Page 179: SIPRI quote is from: SIPRI, *CB Weapons Today,* pp. 314.

Chapter 8

Pages 180–82: General references for this section include: Department of Health and Human Services, op. cit.; Krimsky, op. cit.; OTA, op. cit., pp. 550–51.

Pages 182–83: For a list of P4 labs, see: "Memorandum Opinion and Order," Civil Action No. 84-3542, U.S. District Court for the District of Columbia, May 31, 1985, p. 6. For a good description of the changes in the NIH guidelines, see: *Environment* (July/August 1982), p. 12.

Page 183: First Sinsheimer quote is from: Interview by Piller, December 13, 1982. Second Sinsheimer and King quotes are from DHHS, *Recombinant DNA Research,* vol. 7 (December 1982), pp. 649 and 720 respectively.

Page 184: Nader comments are from: Keynote address at Creating a Public Agenda for Biotechnology, conference sponsored by the Committee for Responsible Genetics, Washington, D.C., November 13, 1987. For a description of Kennedy and Cline violations, see: Jeremy Cherfas, *Man Made Life* (1982), pp. 135–37.

Pages 185–87: For ice-minus background, see, for example: *New York Times,* February 27, 1986; and the *Nation* (October 25, 1986), p. 400. Odum letter appears in: *Science* (September 27, 1985); and also see: *Science News* (May 4, 1985), p. 282.

Pages 187–88: Regarding the Price-Anderson Act, see: *New York Times,* March 14, 1986. Rifkin suit is: U.S. District Court for the District of Columbia, Civil Action 86-1590. Bouckaert quote is from: *Wall Street Journal,* March 3, 1986. Wistar and Oregon State incidents are described, respectively, in: *New York Times,* November 11 and 13, 1986. For new biotechnology policy, see: Executive Office of the President, "Coordinated Framework for Regulation of Biotechnology" (1986).

Pages 188–89: Kingsbury quote is from: *New York Times,* November 11, 1986. General Accounting Office report is: GAO, *Biotechnology: Agriculture's Regulatory System Needs Clarification,* March 1986. Cavalieri and King quotes are from, respectively: *New York Times,* June 19 and May 22, 1986. For details on worst-case scenario rule, see: *Washington Post,* April 25, 1986.

Pages 189–90: For DOD order on rDNA guidelines, see: James P. Wade, Jr., acting undersecretary of defense, memorandum, "DOD Research Activities in the Field of Recombinant Deoxyribonucleic Acid (DNA)," April 1, 1981. Barban verified IBC discrepancies in interviews by Piller, October 24, 1983, and February 19, 1987.

Pages 190–91: Main source for Goldstein-Novick proposal: DHHS, *Recombinant DNA Research . . . ,* pp. 824–25 (proposal text and rationale), pp. 783–85 (ACDA position), pp. 788–89 (DOD position), pp. 459–70 (RAC meeting minutes). See also: *Nature* (July 8, 1982), p. 111. Regarding diphtheria toxin, see: *Los Angeles Times,* January 1, 1983; *Bulletin of the Atomic Scientists* (November 1983), p. 20; and RAC minutes of meeting, October 25, 1982.

Pages 191–92: Regarding Warnke's Shiga toxin initiative, see: Statement released by Foundation on Economic Trends (Washington, D.C.), February 6, 1984; and *Washington Post,* February 7, 1984. Regarding Rifkin's subgroup proposal, see: RAC notice of proposed actions, March 28, 1985; and *San Francisco Chronicle,* May 4, 1985.

Pages 192–94: Goldstein quote is from: Interview by Piller, March 31, 1983. King quote is from: *San Jose Mercury News—West,* April 15, 1984, p. 35. Hens Green comment is from: "Memorandum Opinion and Order," Civil Action No. 84-3542, U.S. District Court for the District of Columbia, May 31, 1985, pp. 25–26. Regarding Rifkin's February 1987 victory, see: U.S. District Court for the District of Columbia, Civil Action 86-2436, September 23, 1986; and *New York Times,* February 19, 1987. Open-air testing law is Public Law No. 91-121, November 19, 1969, as amended (50 U.S.C., subsections 1511-1520).

Pages 195–96: Wald quote is from: DHHS, *Recombinant DNA Research. . . ,* p. 701. Krimsky quotes are from: Speech at the College of Marin (California), May 14, 1986. For background on Kingsbury's ties to the biotechnology industry, see: *San Francisco Chronicle,* October 16, 1987. Regarding industry support for biotechnology, see: *Science* (January 17, 1986), p. 242; (June 13, 1986), p. 1361. King quote is from: Interview by Piller, November 4, 1986.

Chapter 9

Pages 197–200: For an overview of the postwar research apparatus, see, for example: Stanton A. Glantz, et al., *DOD Sponsored Research at Stanford,* 1971; Martin L. Perl, *Physics Careers and Education* (1978), pp. 109–22; and *Science* (November 22, 1974), p. 706.

Page 198: Glantz quote is from: Perl, op. cit., p. 113. DOD official is quoted in: U.S. House of Representatives, Committee on Appropriations, "DOD Appropriations for 1956," 1955, p. 260.

Pages 198–99: Glantz quote is from: Perl, op cit., p. 115. For Mansfield Amendment, see: U.S. Senate, Committee on Armed Services, 91st Congress, lst Session, "Hearings on S. 3367 and H.R. 17123," 1970, pp. 118–19, 159. House Committee quote is from: U.S. House, "Report No. 91-552" (accompanying H.R. 14000), 1969, pp. 46–47. Glantz and Albers quote is from: *Science* (November 22, 1974), p. 706.

Pages 200–202: Laird is quoted in: *Science* (February 25, 1972), p. 866. All Glantz and Albers quotes are from: *Science* (November 22, 1974), p. 706. Frosch is quoted in: U.S. Senate, Committee on Armed Services, "Authorizations for Military Procurement, Research and Development, Fiscal Year 1971, and Reserve Strength," part 2, p. 1505. Separate Glantz quote is from: Perl, p. 117.

Pages 202–3: Regarding federal R&D, see: National Science Foundation, *Science Indicators/The 1985 Report,* p. 226. Holdren study is described in: *San Francisco Chronicle,* October 16, 1986. Regarding Thurow's estimates, see: *Discover* (January 1987), p. 94.

Pages 203–4: DOD officials are quoted in: *Science* (November 22, 1974), p. 706. University of Pittsburgh and University of Washington contracts are described in: DOD Forms 1498, Agency Accession No. DA308858, November 25, 1985; and DA302383, March 29, 1985, respectively.

Pages 204–6: Baxter quotes are from: *San Jose Mercury News—West,* April 14, 1984, p. 12. Glantz interview conducted by Piller, January 25, 1984. For a detailed look at MKULTRA, see: Marks, op. cit., and Senate, "Biomedical and Behaviorial. . . ." Regarding Project CHATTER, and for quotes of MKULTRA sources, see: Marks, op. cit., pp. 35–43 and p. 160, respectively.

Pages 207–9: Virginia Polytechnic and Khorana projects are described in: DOD Forms 1498, No. N00014-77-C-0246 (January 7, 1983) and No. N00014-82-K-0668 (May 27, 1983), respectively. Meselson comments are from: Interview by Piller, November 5, 1986. Hicks is quoted in: *Science* (April 25, 1986), p. 443. Drell response and Hicks rejoinder are from: *Science* (June 6, 1986), p. 1183.

Pages 209–11: Novick and Meselson comments are from: Interviews by Piller, November 11 and 5, 1986, respectively. Keyworth comments are from: *Science* (April 6, 1984), p. 9. Weisner, Kemnitzer, and Rice quotes are from: *New York Times,* September 8, 1986. Park quote is from: *New York Times,* June 26, 1983. Atkinson and Kruytbosch views are from: Interviews by Piller, February 9 and 12, 1987, respectively. Glantz quote is from: Perl, op. cit., p. 116.

Pages 212–16: For a detailed overview of NIH peer review procedures, see: NIH, "NIH Public Advisory Groups," January 1985. Newburgh interview was conducted by Piller, September 26, 1986. Information on AIBS was obtained from: Interview of AIBS spokesman Donald Beem by Piller, September 23, 1986; and AIBS (based in Washington, D.C.) descriptive materials. AMRDC information obtained from: AMRDC, "Research Program: Broad Agency Announcement," August, 1986.

Pages 216–18: Noyes comments are from: Interview by Piller, September 24, 1986. ARO background is from: Interview of Tove by Piller, September 18, 1986. List of NRC panelists was obtained from: Interview with Chester McKee, NRC Office of Science and Engineering Personnel, by Yamamoto, February 6, 1987. Dasey, King comments from: Interviews by Piller, February 24, 1987, November 4, 1986, respectively.

Pages 218–19: MEDLINE search was conducted by Kathleen Rañeses (February 28, March 7 and April 2, 1987).

Pages 219–20: Dupree comments are from: *Nature* (September 18, 1986), p. 213. Hicks quotes and a description of the Packard report are from: *Science* (April 25, 1986), p. 443.

Chapter 10

Pages 221–22: Warnke quote is from: *New York Times,* September 27, 1986. For an excellent overview of the strategic arms race, including detailed references, see: Tom Gervasi, *The Myth of Soviet Military Supremacy* (1986), p. 400–02. For Soviet science data, see: NSF, *Science Indicators,* pp. 187, 192. Regarding problems of openness in Soviet research, see: *New York Times,* October 10, 1986; and the *Nation* (June 13, 1987), p. 804.

Pages 222–23: Regarding U.S. estimates of Soviet computer industry, see: *Washington Post,* March 27, 1987. For data and background on U.S. and Soviet international authorship and relative priorities in various scientific fields, see: NSF, *Science Indicators,* pp. 6, 8, 192–95, 215. Ashford comments are from: *New York Times,* October 10, 1986.

Pages 224–25: For all Abrams quotes and a good background on Reagan administration constrictions on information, see: *New York Times Magazine,* September 25, 1983, p. 22. Regarding CIA exemptions to the FOIA, see: *New York Times,* April 17, 1985. Regarding 800,000 improper classifications, see: *New York Times,* April 19, 1983. Reagan executive order on classification is published in: House, *Compilation of Intelligence Laws and Related Laws and Executive Orders of Interest to the National Intelligence Community* (1983), pp. 323–33. Regarding photooptical research censorship, see, for example: *New York Times,* April 9, 1985. GAO censorship estimates are contained in: GAO, "Polygraph and Prepublication Review Policies of Federal Agencies" (GAO/NSIAD-84-134), June 11, 1984, pp. 2, 7.

Pages 226–28: Halperin quote is from: *New York Times,* December 31, 1986. Regarding the CIA's decline under Turner, see: The *Nation* (January 31, 1987), p. 114. Haass and aviation disinformation quotes are from: *Columbia Journalism Review* (January/February 1987), p. 15. Poindexter memo is quoted in: *Washington Post,* October 2, 1986. Shultz, Weintraub quotes are from: *New York Times,* October 3.

Pages 228–31: Reagan is quoted in: Alan Wolfe, *The Rise and Fall of the Soviet Threat* (1984), p. 1. Weinberger quote is from: *New York Times,* April 11, 1984. Gervasi quotes (in order of appearance) are from: Gervasi, op. cit., p. 15 (new weapons); pp. 91, 95 (yields); p. 77 (gaps). Warnke quote is from: *New York Times,* March 31, 1982. For a detailed account of the CIA-DOD conflict on Soviet missiles, see, for example: *New York Times,* July 16, 1986. Regarding spending, see: *San Francisco Chronicle,* March 29, 1986; and Gervasi, op. cit., pp. 212–15. Holzman quote is from: *New York Times,* March 4, 1986. Wolfe quote is from: Wolfe, op. cit., p. 11.

Pages 231–32: For a detailed overview of alleged Soviet violations, see: U.S. Arms Control and Disarmament Agency, "Soviet Noncompliance" (ACDA Publications 120), March 1986. For the contradictory views of various experts, see, for example: *New York Times Book Review,* March 11, 1984, page 3; *Los Angeles Times,* January 19, 1984; and *New York Times,* June 6, 1986. For an analysis of the Stanford report, see: *San Francisco Chronicle,* February 13, 1987. For Joint Chiefs' SALT II position, see: *New York Times,* February 8, 1986.

Pages 232–33: Chain quote is from: *New York Times,* January 10, 1986. *New York Times* editorial appeared January 22, 1987. Regarding U.S. exceeding SALT II limits, see, for example: *New York Times,* November 29, 1986. For a detailed analysis of the controversy over Reagan's reinterpretation of the ABM treaty, see, for example: *New York Times,* February 17, 1987. For critics' comments on the ABM interpretation, see, for example: *New York Times,* February 7 and 26 and March 10, 1987.

Page 234: Wicker's comments are from: *New York Times,* February 11, 1987. Falk's comments are from: Geissler, op. cit., pp. 109, 116. Gervasi quote is from: op. cit., p. 71.

Page 237: Delegate quote is from: *American Society of Microbiology News* (March 1987), p. 124.

SELECTED BIBLIOGRAPHY

UNITED STATES GOVERNMENT DOCUMENTS

Arms Control and Disarmament Agency. *Arms Control and Disarmament Agreements: Texts and Histories of Negotiations,* 1982.

Congressional Office of Technology Assessment. *Commercial Biotechnology: An International Analysis* (OTA-BA-218), January 1984.

Department of Defense. "Annual Report on Chemical Warfare-Biological Defense Program Obligations," fiscal years 1976–85.

———. "Biological Defense Program: Report to the Committee on Appropriations, House of Representatives," May 1986.

———. Defense Intelligence Agency. "Soviet Biological Warfare Threat" (DST-161F-057-86), 1986.

———. "Soviet Military Power," April 1984.

Department of Health and Human Services. National Institutes of Health. "Guidelines for Research Involving Recombinant DNA Molecules," May 1986.

———. *Recombinant DNA Research,* vol. 7, December 1982.

Executive Office of the President. Office of Science and Technology Policy. "Coordinated Framework for Regulation of Biotechnology," June 26, 1986.

General Accounting Office. *Biotechnology: Agriculture's Regulatory System Needs Clarification* (GAO/RCED-86-39BR), March 1986.

House of Representatives. Committee on Government Operations. Conservation and Natural Resources Subcommittee. (Hearings.) "Environmental Dangers of Open-Air Testing of Lethal Chemicals," May 20–21, 1969.

———. Permanent Select Committee on Intelligence. *Compilation of Intelligence Laws and Related Laws and Executive Orders of Interest to the National Intelligence Community*. April 1983.

National Science Foundation. "Federal Funds for Research and Development," fiscal years 1973–87.

———. *Science Indicators/The 1985 Report.*

Senate, Committee on Foreign Relations. (Hearings.) "The Geneva Protocol of 1925," March 5, 16, 18, 19, 22, 26, 1971.

———. Committee on Human Resources, Subcommittee on Health and Scientific Research. (Hearings.) "Biological Testing Involving Human Subjects by the Department of Defense, 1977," March 8 and May 23, 1977.

———. Committee on Labor and Public Welfare, Subcommittee on Health; and Committee on the Judiciary, Subcommittee on Administrative Practice and Procedure. (Joint hearings.) "Biomedical and Behavioral Research, 1975," September 10 and 12 and November 7, 1975.

———. Select Committee to Study Governmental Operations with Respect to Intelligence Activities. (Hearings.) "Unauthorized Storage of Toxic Agents," September 16–18, 1985.

BOOKS

Cherfas, Jeremy. *Man-Made Life.* New York: Pantheon, 1982.

Burrows, William E. *Deep Black: Space Espionage and National Security*. New York: Random House, 1986.

Geissler, Erhard, ed. *Biological and Toxin Weapons Today*. London: SIPRI/Oxford University Press, 1986.

Gervasi, Tom. *The Myth of Soviet Military Supremacy*. New York: Harper & Row, 1986.

Glantz, Stanton A., et al. *DOD Sponsored Research at Stanford, Vol. I, Two Perceptions: The Investigator's and the Sponsor's, Vol. II, Its Impact on the University*. Palo Alto, Calif.: Stanford University Workshops on Political and Social Issues, 1971.

Harris, Robert, and Jeremy Paxman. *A Higher Form of Killing.* New York: Hill and Wang, 1982.

Krimsky, Sheldon. *Genetic Alchemy.* Cambridge: MIT Press, 1982.

Marks, John. *The Search for the "Manchurian Candidate."* New York: McGraw-Hill, 1980.

Murphy, Sean; Alastair Hay; and Steven Rose. *No Fire No Thunder.* New York: Monthly Review Press, 1984.

Perl, Martin L. *Physics Careers, Employment and Education.* New York: American Institute of Physics, 1978.

Stockholm International Peace Research Institute. *Arms Control: A Survey and Appraisal of Multilateral Agreements.* London: Taylor and Francis, 1978.

———. *World Armaments and Disarmament: SIPRI Yearbooks.* London: Taylor and Francis, 1976, 1982, 1983, 1984, 1985, 1986.

———. *The Problem of Chemical and Biological Warfare.* Stockholm: Almqvist and Wiksell.
Vol. I, The Rise of CB Weapons. 1971.
Vol. II, CB Weapons Today. 1973.
Vol. III, CBW and the Law of War. 1973.
Vol. IV, CB Disarmament Negotiations, 1920–1970. 1971.
Vol. V, The Prevention of CBW. 1971.
Vol. VI, Technical Aspects of Early Warning and Verification. 1975.

Wolfe, Alan. *The Rise and Fall of the Soviet Threat.* Boston: South End Press, 1984.

Zilinskas, Raymond A. "Managing the International Consequences of Recombinant DNA Research." Ph.D. dissertation. Los Angeles: University of Southern California, 1981.

INDEX

ABOUT THE AUTHORS

Charles Piller is an investigative journalist whose articles on biological warfare, science, labor and international affairs have appeared in magazines and newspapers across the country, including the *Nation, Los Angeles Times, Philadelphia Inquirer,* and *Baltimore Sun*. He has served as a consultant to the U.S. Senate for an oversight investigation of military research activities and to CBS News and local television producers for broadcasts on chemical and biological weapons. Piller has lectured on these issues at the University of California at Berkeley since 1985.

Keith R. Yamamoto, professor and vice-chairman in the Department of Biochemistry and Biophysics at the University of California, San Francisco, obtained his Ph.D. in 1973 at Princeton University. His lab at UCSF pursues basic molecular biological research involving hormonal control of gene expression. He currently chairs the molecular biology study section for the National Institutes of Health and served on the genetic biology review panel for the National Science Foundation.